A FORMULARY OF ADHESIVES AND OTHER SEALANTS

Compiled by

Michael and Irene Ash

Chemical Publishing Co., Inc.
New York, N.Y.

© **1987**

Chemical Publishing Co., Inc.

ISBN 0-8206-0297-3
ISBN 0-8206-0366-X (pbk)

Printed in the United States of America

PREFACE

Through the cooperation and contribution of major adhesive raw material manufacturers, we are pleased to present a formulary devoted entirely to a variety of adhesives and sealants. The purpose of this compilation is not to educate the reader as to the physical and chemical nature of adhesive ingredients but to make available a catalog of formulas reflecting the current technology in the adhesives industry.

The formulas herein should be considered starting-point preparations and thereby used as the basis for further experimentation in the development of new adhesive products.

The book is organized so that each chapter deals with major adhesive applications. Within each chapter the formulas are further subdivided according to their specific use and chemical composition.

Unless otherwise specified, all formulas have the quantities of ingredients given in parts by weight. A list of abbreviations that are used throughout the formulary is included. All constituents appearing by their tradename are printed in boldface type, and the manufacturers' names and addresses appear after the list of alphabetized tradenames in the appendix.

ABBREVIATIONS

@	at
approx.	approximately
aq.	aqueous
ASTM	American Society for Testing and Materials
Bé.	Baume
b.p.	boiling point
C.	degrees Centigrade
ca.	circa
cc.	cubic centimeter
cm.	centimeter
conc.	concentrated
cP	centipoise(s)
cps	centipoise(s)
cs.	centistoke(s)
F.	degrees Fahrenheit
fl oz.	fluid ounce
ft.	foot
g.	gram
gal.	gallon
h.	hour
ht.	height
in.	inch
l.	liter
m.	meter
M.	mole
max.	maximum
med.	medium
min.	minute
min.	minimum
ml.	milliliter
mm.	millimeter
m.p.	melting point
m.w.	molecular weight
N.	newton
neut.	neutralized
NF	National Formulary

CONTRIBUTORS

American Cyanamid Co.
Wayne, NJ 07470

Amoco Chemical Corp.
200 E. Randolph Drive
Chicago, IL 60601

Arizona Chemical Co.
Wayne, NJ 07470

BASF Wyandotte Corp.
Industrial Chemicals Group
1609 Biddle Ave.
Wyandotte, MI 48192

H. Bennett
4747 Collins Ave.
Miami Beach, FL 33140

E.I. DuPont de Nemours Co.
Wilmington, DE 19898

Genstar Stone Products Co.
Executive Plaza IV
Hunt Valley, MD 21031

Goodyear Tire & Rubber Co.
1144 East Market St.
Akron, OH 44316

Hercules Inc.
910 Market St.
Wilmington, DE 19899

Merck & Co. Inc./Chemical Div.
126 E. Lincoln Ave.
Rahway, NJ 07065

Monsanto Co.
800 No. Lindbergh Blvd.
St. Louis, MO 63166

Neville Chemical Co.
Neville Island
Pittsburgh, PA 15225

Pacific Scott Bader Inc.
1148 Harbour Way S.
Richmond, CA 94804

Petrolite Corp./Bareco Div.
6910 East 14 St.
Tulsa, OK 74115

Reichhold Chemicals Inc.
Resins and Binders Div.
Pensacola, FL 32596

Rohm & Haas Co.
Independence Mall West
Philadelphia, PA 19105

Shell Chemical Co.
One Shell Plaza
Houston, TX 77002

A.E. Staley Mfg. Co.
Box 151
Decatur, IL 62525

Thompson, Weinman & Co.
PO Box 130
Cartersville, GA 30120

R.T. Vanderbilt
30 Winfield St.
Norwalk, CT 06855

Uniroyal Chemical
Div. Uniroyal Inc.
Elm St.
Naugatuck, CT 06770

TABLE OF CONTENTS

Chapter I

ADHESIVES FOR PAPER

Pressure-Sensitive Labels

Formula No. 1

(Rubber/Resin)

Kraton®	100
Wingtack® 95	140
Shellflex® 371	40
Irganox® 1010	2

No. 2
(Permanent, Rubber/Resin)

SBR-1011	100.0
Super Sta-Tac® 0-80	150.0
Nonstaining Antioxidant	1.5
Toluene	200.0
Hexane	200.0

Properties:

Solids Content	38 ± 1%
Visc. (Brookfield)	1800–2200 cps

No. 3

(Polyvinyl Isobutyl Ether)

Lutonal IC 125	30.0
Lutonal I 60	40.0
Lutonal I 30	10.0
Vulcanox® ZKF	0.5
Mineral Spirit (b.p. 65–95 C)	≈ 150.0

No. 4

(Vinyl-Isobutyl Ether)

Lutonal IC 30	30.0
Lutonal I 60	60.0
Lutonal IC 125	30.0
Vulcanox® ZKF	1.0
Mineral Spirit (b.p. 65–95 C)	≈ 120.0

No. 5

(Water-Soluble, Vinyl Ether/Resin)

Acronal 880 D	50
Acronal 500 D	30
Lutonal M 40 (≈50% in water)	20
Lutensol AP 6	1
Toluene	5

No. 6

(Peelable, Acrylic Resin)

Acronal 4 D	90
Acronal 80 D	10

No. 7

(Peelable, Acrylic Resin)

Acronal 4 L ($\approx$ 25% in acetone/mineral spirit)	90
Acronal 4 L ($\approx$50% in ethyl acetate)	10

No. 8

(Peelable, Acrylic Resin)

Acronal 4 L ($\approx$ 25% in acetone/mineral spirit)	90
Staybelite® Ester 10 (30% sol'n. in toluene)	10

No. 9

(Peelable, Acrylic Resin)

Acronal 103 L	100.0
Basonat A 270	0.5

No. 10

(Removable, Acrylic Resin)

Acronal 4 D	100
Palatinol® AH	20

No. 11

(Water Removable, Acrylic Resin)

Acronal 500 D	30
Acronal 880 D	50
Lutonal M 40 (50% in water)	20
Lutensol® AP 6	1

Note:
This adhesive loses its adhesion on short contact with water. It is re-emulsified, but not dissolved.

Lutonal M 40, 50% in water, can be more readily incorporated after it has been diluted with 20–40 parts water.   Temperatures above 28 C should be avoided because **Lutonal M 40** may precipitate under such conditions.

No. 12

(Permanent, Acrylic Resin)

Acronal 3 L ($\approx$ 25% in acetone/mineral spirit)	80.0
Carbigen® K 90 (50% sol'n. in toluene)	20.0
Desmodur® L (75% in ethyl acetate)	0.5

No. 13

(Permanent, Acrylic Resin)

Acronal V 205 (pH 7–8)	90
Lutonal I 65 D	10

No. 14

(Permanent, Acrylic Resin)

Acronal V 205	90
Acronal 7 D	10

No. 15

(Permanent, Acrylic Resin)

Acronal 85 D	90
Acronal 7 D	10

No. 16

(Permanent, Acrylic Resin)

Acronal 80 D	60
Snowhite® 301 CF	40

No. 17

(Polyisobutylene)

Oppanol B 15	85
Oppanol B 100	15
Cellolyn® 21	10
Mineral Spirit (b.p. 65–95 C)	≈ 150

Note:

Since the **Oppanol B** types have more or less pronounced creep at room temperature, special attention must be paid to the adhesive properties and the strike-through behavior in the trials.

Despite their limited compatibility with one another, the **Oppanol B** types can be used in combination with rubber. The creep at room temperature can thus be reduced.

No. 18

(Removable, Polyisobutylene)

Oppanol B 3	30
Oppanol B 120	15
Mineral Spirit (b.p. 65–95 C)	90

Pressure-Sensitive Labels for Deep-Frozen Articles

Formula No. 1

(Vinyl Isobutyl Ether/Resin)

Acronal V 205	87
Acronal 7 D	8
Plastilit 3060	5

No. 2

(Solvent-Based, Phenol)

Lutonal I 60	60.0
Lutonal IC 125	30.0
Vulcanox® ZKF	0.9
Mineral Spirit (b.p. 65–95 C)	≈ 120.0

No. 3

(Vinyl Methyl Ether/Resin)

Acronal V 205	70
Lutonal M 40 (50% in water)	30
Water	40
Toluene	5

Note:
This formulation has good adhesion to moist surfaces.

		No. 4	No. 5	No. 6	No. 7
			(Rubber/Resin)		
Kraton® D-1197		100	100.0	100	100
Wingtack® 76		75	—	—	—
Super Sta-Tac® 80		—	62.5	—	—
Escorez® 1310		—	—	50	—
Wingtack® 95		—	—	—	50
Wingtack® 10		50	62.5	75	100
Antioxidant	Target Requirements*	2–5	2–5	2–5	2–5
Properties:					
Tg (calculated) °C	—		−21	−24	−26	−25
Dead Weight Tack, (stainless steel probe)					
−18 C, g	>280	390	420	450	470
Rolling Ball Tack, cm	<2.0	1.2	1.0	1.1	1.0
Polyken® Probe Tack @ R.T., kg	>0.5	0.6	0.6	0.5	0.6
Loop Tack**, N/m	>340	540	620	580	700
180° Peel Adhesion, N/m	>420	580	620	660	770
Melt Visc. at 177 C, Pa·s	<50	43	41	43	23

---

* As measured on commercial low temperature labels.

** To steel, 25 × 25 mm contact area.

Label Adhesives

(Rubber/Resin)

Kraton® 1102	100.0
Super Sta-Tac® 80	90.0
Super Sta-Tac® 100	40.0
Nirez® 1010	20.0
Antioxidant	1.0

Properties:
Visc. (Brookfield Thermosel) @ 350 F—58,000 cps
30 s Chromecote Adhesion to Glass—Paper failure upon removal
($3/4$ mil coating thickness)
Bleeding Test—72 h @ 160 F—Adhesion exhibits very slight bleeding

Paper-to-Glass Nonwater-Resistant
Label Adhesive

(Water-Based, Dextrin)

Water	15.1
Stadex Dextrin No. 126	30.2

Heat to 190 F. Hold at 190 F for 10 min.
Add:

Corn Syrup (43 Bé)	45.3
Ethylene Glycol	8.4

Heat to 160 F. Mix well. Cool and drum.

Properties:
Solids (refractometer)	73–75%
pH	≈ 3.5
Visc. (Brookfield) @ 80 F (#6 spindle, 20 rpm)	6,000–10,000 cps

Low-Temperature Labels

	Formula No. 1	No. 2
	(Rubber/Resin)	
Kraton® D-1107	100	—
Kraton® DX-1112	—	100
Escorez® 1310	50	50
Wingtack® 10	75	75
Antioxidant	2–5	2–5

Hot-Melt, Pressure-Sensitive Labels

Formula No. 1

(Styrene–Isoprene/Resin)

Zonarez 7115	40.0
Styrene/Isoprene Block Copolymer	60.0
Antioxidant	0.5

Properties:
Kraft Backing

Quick Stick (PSTC-5)	fiber tear
Shear Adhesion*	1035 min.
Rolling Ball Tack (PSTC-6)	0 in.
Visc. @ 300 F	75,000 cps
@ 350 F	20,750 cps
@ 390 F	8,600 cps

Note:
This formulation, when applied to standard paper stock, will produce paper fiber tear before stripping from substrate.

---

* Modified PSTC-7; 20° shear test using 100 g weight on $^{1}/_{2}$ in.^{2} test area.

No. 2

(Rubber/Resin)

Kraton® DX 112	100
Escorez® 1310	50
Wingtack® 10	75
Antioxidant	2

No. 3

(Rubber/Resin)

Kraton® 1107	80
Kraton® 1102	20
Zonester® 85	150
Arizona DR 24	50
Tufflo® 6056	75
Vanox 13	1

Properties:

Visc. @ 350 F	3500 cps
180° Peel Adhesion (PSTC-1)	100 oz/in.
Adhesion* to:	
Nitrocellulose–Wax Coated Cellophane	Excellent
Wetting agent–Polyvinylidene Chloride Coated Cellophane	Very good
Polyethylene Terephthalate (Polyester)	Excellent
Surylyn Ionomer Resin	Excellent
Glass	Excellent
Polyethylene	Excellent

---

* Excellent bonds are permanent and tear paper backing on removal. Fair bonds are adequately adherent removable bonds.

No 4

(Rubber/Resin)

Kraton® 1107	80
Kraton® 1102	20
Zonester® 100	150
Arizona DR 24	50
Tufflo® 6056	75
Vanox 13	1

Properties:

Visc. @ 350 F	6000 cps
180° Peel Adhesion (PSTC-1)	95 oz/in.
Adhesion* to:	
Nitrocellulose–Wax Coated Cellophane	Good
Wetting agent–Polyvinylidene Chloride Coated Cellophane	Fair
Polyethylene Terephthalate (Polyester)	Excellent
Surylyn N Ionomer Resin	Excellent
Glass	Excellent
Polyethylene	Good

---

* Excellent bonds are permanent and tear paper backing on removal. Fair bonds are adequately adherent removable bonds.

No. 5

(Removable, Rubber/Resin)

Kraton® 1107	33.0
Sta-Tac®-R	32.5
Shell Flex 371	33.0
Irganox® 1010	1.0
Butyl Zimate	0.5

Properties:

Visc.	@ 300 F	12,312.5 cps
	@ 350 F	5,687.5 cps

Following properties were collected from an 1.0 mil coating of H-1048 on Mylar (Du Pont type S, 2.0 mil).

Rolling Ball Tack	1 $1/_{16}$ in.
180° Peel	362 g/in. of width
Quick Stick (Loop Test)	544 g/in.
0° Shear (2 kg/in.^{2})	12.5 min.

Following properties were collected from an 0.5–0.75 mil coating of H-1048 on 60-pound Gloss label stock.

180° Peel (to Stainless Steel)	317.5 g/in. of width

Aging at 70 C:

Bleed Test: Samples of H-1048 (label stock to release liner and label stock bonded to glass) exhibited slight "bleeding" after 14 days at temperature. The sample bonded to glass was easily removed (peeled clean) after the 14-day exposure.

	No. 6	No. 7
	(Rubber/Resin)	
Zonester® 85	75	10
Kraton® 1102	100	100
Tufflo® 6056	60	60
Vanox 13	1	1

Properties:

Visc. @ 350 F	19,000 cps	15,000 cps

Note:

Good adhesion to the following common packaging substrates: cellophane, mylar, Kraft paper, glass, aluminum foil, and polyethylene.

No. 8

(Permanent, Rubber/Resin)

Kraton® 1102	39.6
Super Sta-Tac®-80	36.0
Super Sta-Tac®-100	16.0
Nirez® 1010	8.0
Antioxidant	0.4

Procedure:

Prepare in Baker-Perkins dispersion blade mixer with steam heat of 225 F.

Properties:

Visc. @ 350 F	58–60,000 cps
30-s Adhesion to Glass	Excellent substrate failure

(sample coated on chrome coat label stock at $3/4$ mils)

"Bleeding" Test @ 180 F	72 h

The above system exhibits slight to moderate bleeding during the test cycle.

No. 9

(Ethylene Vinyl Acetate/Resin)

Be Square® 195	30
Elvax 250	50
Dymerex	20

Properties:

Visc.—Initial, ASTM D-3236　@ 250 F (121 C)　　　91,600 cps
　　　　　　　　　　　　　@ 275 F (135 C)　　　54,400 cps
　　　　　　　　　　　　　@ 300 F (149 C)　　　33,300 cps
　　　　　　　　　　　　　@ 325 F (163 C)　　　22,000 cps
　　　　　　　　　　　　　@ 350 F (177 C)　　　15,000 cps
Visc.—Aged, after 32 h @ 300 F (149 C), ASTM D-3236
　　　　　　　　　　　　　@ 300 F (149 C)　　　38,000 cps
Aged Appearance—
　　After 72 h @ 300 F (149 C)　　Color change and slight sludge
Penetration—ASTM D-1321 @ 100 F (38 C)　　　　　5.7
Softening Point—ASTM E-28　　　　　　　　193 F (89 C)

Note:

Very fast set-up time.　Adheres well to foil and medium density polyethylene and most other packaging films.

Uses:

Adhesive coating to foil label for closure of individual portions of honey, jelly, jams, etc.　Provides good adhesive between the foil and medium density polyethylene but is peelable.

No. 10

(Ethylene-Vinyl Acetate/Resin)

Ultraflex®, Victory®, or Be Square® 175	30
Elvax 250	50
Staybelite®	20

Properties:

Visc.—Initial, ASTM D-3236　@ 250 F (121 C)　　　5560 cps
　　　　　　　　　　　　　@ 275 F (135 C)　　　3500 cps
　　　　　　　　　　　　　@ 300 F (149 C)　　　2350 cps
　　　　　　　　　　　　　@ 325 F (163 C)　　　1650 cps
　　　　　　　　　　　　　@ 350 F (177 C)　　　1200 cps
Visc.—Aged, after 32 h @ 300 F (149 C), ASTM D-3236
　　　　　　　　　　　　　@ 300 F (149 C)　　　3120 cps
Aged Appearance—
　　After 72 h @ 300 F (149 C)　　　Slight color change
Softening Point—ASTM E-28　　　　　　　154 F (68 C)
Penetration—ASTM D-1321 @ 100 F (38 C)　　　　22

The above data obtained with an **Ultraflex®** based HMA. The use of **Victory®** changes the softening point to 163 F (73 C) and penetration to 18. The use of **Be Square® 175** changes the softening point to 178 F (81 C) and the penetration to 14.

Note:
Nonblocking when gravure applied. Adheres to most packaging films including polypropylene.

Uses:
Particularly useful for end and top labels on polypropylene.

No. 11

(Ethylene Vinyl Acetate/Resin)

Elvax 250	25.0
Ultraflex®	32.5
Staybelite®	20.0
Low molecular wt., oxidized polyethylene	
(Such as **AC629, Epolene E-10,** etc.)	20.0
Armid C	2.5

No. 12

(Ethylene Vinyl Acetate/Resin)

Aluminum Stearate	7.5
Staybelite® Ester #10	25.0
Ceraweld	80.5

No. 13

(Ethylene Vinyl Acetate/Resin)

Elvax 250	50
Dymerex	20
Be Square® 190/195	30

No. 14

(Ethylene Vinyl Acetate/Resin)

Elvax 250	30
Ultraflex®	50
Staybelite®	20

No. 15

(Ethylene Vinyl Acetate/Resin)

Zonarez 7115	33.3
EVA (28% VA-6 MI)	33.3
Microcrystalline Wax (m.p. 180 F)	33.3

Properties:

Sealing Temp.	350 F
Visc. @ 350 F	5560 cps
@ 370 F	4200 cps
@ 380 F	3700 cps
Adhesion:	
Kraft to Kraft	29 oz/in.
Kraft to PE	35 oz/in.
Kraft to PP	9 oz/in.
Kraft to Aluminum foil	18 oz/in.

Note:

This is a low-cost, low-viscosity adhesive which exhibits good adhesion to films and foil.

No. 16

(Ethylene-Vinyl Acetate/Resin)

Ultraflex®	32.5
Elvax 250	25.0
Staybelite®	20.0
Cardipol® LP O-25	20.0
Armid C	2.5

Note:

Excellent adhesion to polymeric films. Excellent adhesion to polypropylene film at freezer temperature.

Uses:

Bread end labels, etc.

	No. 17	No. 18	No. 19	No. 20	No. 21	No. 22	No. 23
	(General Purpose, Low Cost)			(UV-Resistant)	(Soft)	(High Shear Strength)	(High Tack)
	(Ethylene Vinyl Acetate/Resin)						
Elvax EP 170-PS	30	26	24	30	25	26	30
Elvax 40-P	—	—	—	—	10	—	—
Piccovar® L-30S	—	—	—	32	—	—	—
Piccovar® L-60	—	—	—	35	—	—	55
Arizona DR-24	65	59	28	—	50	49	—
Zonester® 100	—	—	28	—	—	—	—
Ditridecyl Phthalate	5	—	—	3	—	—	—
Circosol 4240	—	15	20	—	15	25	15
Shellflex® 371	—	—	—	—	—	—	1.0
Weston 399B	1.0	1.0	1.0	—	1.0	1.0	1.0
BHT	—	—	—	0.2	—	—	—
Eastman DOBP	—	—	—	2.0	—	—	—
Properties:							
Visc. @ 149 C (300 F), mPa·s (cP)	42,000	19,000	17,000	51,000	32,000	26,000	30,000

No. 24

(Butadiene-Styrene/Resin)

Ameripol 1011	100
Staybelite® Ester 10	75
Antioxidant 2246	2
Toluene	1003

Hot-Melt, Pressure-Sensitive Labels for Polyethylene and Aluminum Foils

Formula No. 1

(Ethylene Vinyl Acetate/Resin)

Zonarez B-125	20.0
EVA (28% VA–25 MI)	50.0
Microcrystalline Wax (m.p. 190–195 F)	30.0

Properties:

Sealing Temp.	350 F
Visc. @ 350 F	11,870 cps
@ 390 F	6,760 cps

Hot-Melt Labels

Formula No. 1

(Resin)

Polyethylene Resin, M_{vis} 8000	68.0
Piccolyte®	26.5
Polybutene 32	5.0
Piccolastic® A-50	0.5

Note:

This formulation is activated by heating to about 465 F. The polybutene acts as a plasticizer and is particularly effective in improving low temperature flexibility.

No. 2

(Stearate-Gel)

Victory®	80.5
Aluminum Stearate	7.5
Staybelite® Ester 10	12.0

Properties:
Visc.—ASTM D-3236 @ 250 F (121 C)   ≈ 25,000 cps

Note:

Excellent hold-out on paper. Gel prevents slippage of labels while hot. Adheres to paper, board, cellophane, glass, polyethylene, and aluminum foil. Poor adhesion to polypropylene. Must be starch dusted to prevent blocking.

Folding-Carton Sealing Adhesive

(Vinyl Acetate Ethylene)

Airflex 300	79.80
Benzoflex 9-88	9.40
Trichloroethane	4.70
Surfynol® D-201	0.20
Water	5.90

Case-Sealing Adhesive

(Water-Based)

Water	57.0
Stadex 79	35.2
Borax (10 mole)	6.3
Urea	1.0
Defoamer	0.1
Preservative	0.1

Cook 20 min. @ 195 F. Cool below 140 F. Add:

50% Sodium Hydroxide Sol'n.	0.3

Mix and cool. Dilute, if required.

Properties:
  Visc. (Brookfield) @ 80 F   ≈ 1,000 cps
  Refractometer Solids   39–41%

Hot-Melt Carton and Case-Sealing Adhesives

Formula No. 1

(Ethylene Vinyl Acetate/Resin)

Be Square® 195	30
Elvax 250	45
Staybelite® Ester 10	25

Properties:
  Visc.—Initial, ASTM D-3236   @ 250 F (121 C)   47,000 cps
                                @ 275 F (135 C)   29,000 cps
                                @ 300 F (149 C)   18,500 cps
                                @ 325 F (163 C)   12,300 cps
                                @ 350 F (177 C)   8,500 cps
  Visc.—Aged, after 32 h @ 300 F (149 C), ASTM D-3236
                                @ 300 F (149 C)   20,000 cps
  Aged Appearance—
    After 72 h @ 300 F (149 C)   Color change and slight sludge
  Penetration @ 100 F (38 C)   5.5
  Softening Point—ASTM E-28   192 F (89 C)

Note:
Excellent hot tack. Very fast set-up time. Nonblocking. Can be extruded and pelletized. Adheres to cellophane, paper, aluminum, polyethylene, styrene, and many other packaging films. Adhesion to polypropylene is poor.

Uses:
Preapplied heat-seal carbon adhesive. May also be used as carbon adhesive on the package closing machine.

No. 2

(Ethylene Vinyl Acetate/Resin)

Zonester® 85	30.0
Elvax 210	15.0
Elvax 240	20.0
Shellwax 700	28.0
Epolene N-14	5.0
Kraton® 1102	2.0
Butylated Hydroxy Toluene	0.1

Properties:

Visc. (@ 350 F)	1830 cps
10 min. Bond Shear Temp. (100 g/in.)	82 C
Sealing Temp. Range (Du Pont Spring Test)	225–250 F
T-Peel Strength (Kraft to Kraft)	23 oz/in. width
Breaking Strength	415 psi
Elongation at Break	500%

No. 3

(Ethylene Vinyl Acetate/Resin)

Zonester® 100	48.0
Elvax 250	24.0
Epolene N-14	8.0
Shellwax 300	20.0
Butylated Hydroxy Toluene	0.1

Properties:

Visc. (@ 350 F)	1300 cps
10 min. Bond Shear Temp. (100 g/in.)	77 C
Sealing Temp. Range (Du Pont Spring Test)	200–250 F
T-Peel Strength (Kraft to Kraft)	27 oz/in. width
Breaking Strength	550 psi
Elongation at Break	320%

	No. 4	No. 5
	(Ethylene Vinyl Acetate)	
Piccotex® LC	16.0	35.0
EVA (28% VAc 3 MI)	57.0	—
EVA (18% VAc 2.5 MI)	—	25.0
Paraffin Wax (m.p. 155 F)	27.0	—
Microcrystalline Wax (m.p. 180 F)	—	25.0
Polyethylene (AC-8)	—	15.0
Antioxidant (BHT)	0.1	0.2

No. 6

(Ethylene Vinyl Acetate/Resin)

Piccotex® 100	15.0
Elvax 260	12.5
Elvax 220	12.5
Paraffin (≈ m.p. 155 F)	50.0
Micro Wax (≈ m.p. 180 F)	10.0
Tenox BHT	0.1
Armoslip E	0.2

No. 7

(Resin)

Nirez® 1115	49.5
NA-601-04	49.5
Antioxidant	1.0

Procedure:
Melt **NA-601-04**, resin, and antioxidant at 325–350 F, mix until system is uniform.

Properties:
Visc.	@ 300 F	1000–1200 cps
	@ 350 F	500–700 cps

No. 8

(Vinyl Acetate/Resin)

Elvax 150	25.0
Nirez® 1115	54.5
Paraflint H-1	20.0
Antioxidant	0.5

Procedure:
Melt the **Elvax** and resin at 300–350 F, mix with adequate sheer until smooth. Add antioxidant and wax, stir until system is uniform.
Properties:

Visc.	@ 300 F	1900–2000 cps
	@ 350 F	800–900 cps

No. 9

(Ethylene Vinyl Acetate/Resin)

Elvax 250	90.0
Sta-Tac®-R	70.0
Wax (m.p. 150 F)	30.0
Antioxidant	1.0

Properties:
Visc. (Brookfield Thermosel) @ 300 F   44,000–46,000 cps

System coated on carton stock. Initial adhesion good, substrate failure. Low temperature adhesion good, substrate failure. (Test: 45 min. @ –20 F)

No. 10

(Polyethylene/Resin)

Zonarez B-115	25.0
EVA (28% VA–6 MI)	20.0
Epolene C-10	30.0
Microcrystalline Wax (m.p. 180 F)	25.0

Properties:
Sealing Temp.   350 F

No. 11

(Polyethylene/Resin)

Zonarez B-115	30.0
Epolene C-10	48.0
Epolene C-13	22.0

Properties:

Sealing Temp.	350 F
Visc. @ 300 F	9700 cps
@ 350 F	4400 cps
@ 390 F	1303 cps
Adhesion (Kraft to Kraft)	25 oz/in.

Hot-Melt Adhesive for Frozen-Food Cartons

Formula No. 1

(Ethylene Vinyl Acetate/Resin)

Zonarez B-115	20.0
EVA (28% VA-400 MI)	20.0
EVA (28% VA-25 MI)	25.0
Microcrystalline Wax (m.p. 165 F)	35.0

Properties:

Sealing Temp.	350 F
Tensile Strength	384 psi
Visc. @ 340 F	6660 cps
@ 350 F	5800 cps
@ 380 F	3920 cps
Adhesion: Kraft to Kraft	19 oz/in.
Kraft to Chipboard	51 oz/in.

No. 2

(Ethylene Vinyl Acetate/Resin)

Comer EVA-501	20
Comer EVA-505	25
Nirez® 2040	20
Ultraflex®	35

No. 3

(Ethylene Vinyl Acetate/Resin)

Ultraflex®	35
Elvax 210	20
Elvax 250	25
Zonarez 7115	20

Properties:

Visc.—Initial, ASTM D-3236	@ 250 F (121 C)	15,900 cps
	@ 275 F (135 C)	9,800 cps
	@ 300 F (149 C)	6,500 cps
	@ 325 F (163 C)	4,500 cps
	@ 350 F (177 C)	320 cps
Visc.—Aged, after 32 h @ 300 F (149 C), ASTM D-3236		
	@ 300 F (149 C)	6,500 cps
Aged Appearance—		
After 72 h @ 300 F (149 C)		Slight color change
Penetration—ASTM D-1321 @ 100 F (38 C)		20.5
Softening Point—ASTM E-28		157 F (69 C)

Note:

Excellent hot tack, odor, and color. Excellent adhesive at freezer temperatures. Excellent adhesion to a variety of packaging films.

Uses:

Can be applied over most wax based carton coatings providing excellent seal.

Box-Making Adhesive for Small Paper Containers

(Ethylene Vinyl Acetate/Resin)

EVA	40.0
Tackifying Resin	45.0
Wax	15.0
Antioxidant	0.5

Note:

When bonding strength and cold resistance are particularly required. Typical tackifying resins are rosin polyol ester, its hydrogenated materials, and terpene resins used singularly or in combination. Waxes are paraffin wax with melting point at about 145 F and microcrystalline waxes with melting point at about 175 F. are also used either singularly or in combination. For antioxidant, combinations of 4,4'-methylene bis (2,6-di-*tert*-butyl phenol) and tetrakis (methylene 3,3',5'-di-*tert*-butyl-4'-hydroxy phenyl propionate) methane.

Corrugated Cardboard Box Adhesive

(Ethylene Vinyl Acetate/Resin)

EVA	40
Tackifying Resin	45
Microcrystalline Wax	15

The tackifying resins are the same as those described above.

Hot-Melt Packaging Adhesives

	Formula No. 1	No. 2	No. 3	No. 4	No. 5	No. 6	No. 7	No. 8
	(Polyethylene/Resin)							
Petrothene NA593	40	40	40	40	40	40	40	40
Wingtack® 95	40	40	40	40	40	40	40	40
Microwax (m.p. 185 F)	20	20	20	20	—	—	15	15
BHT Antioxidant	1	1	1	1	1	1	1	1
L-14 Polybutene	—	20	40	—	—	—	25	25
H-100 Polybutene	—	—	—	50	—	—	—	—
H-300 Polybutene	—	—	—	—	20	—	—	—
H-1500 Polybutene	—	—	—	—	—	—	5	—
H-1900 Polybutene	—	—	—	—	—	20	—	5
Properties:								
Visc., cps @ 300 F	13,500	4,700	3,500	3,380	31,625	42,188	4,850	5,250
Softening Point, F	212	208	206	208	221	221	210	210
Tensile Strength, psi	69.7	176.2	166.7	157.4	291	274	144	172
Elongation, %	0	40	40	40	288	188	42	32
Shore A Hardness	94	57	51	50	65	67	37	40
Needle Penetration	5	23	29	31	11	11	24	25
Test Results								
(180° Peel Adhesion, oz/in. width)								
Natural Kraft Paper (60 lb/ream)	19.7	23.1	45.1	18.8	20.8	33.6	16.9	14.7

	No. 1	No. 2	No. 3	No. 4	No. 5	No. 6	No. 7	No. 8
Polyethylene Film (untreated)	15.0	15.3	18.0	21.8	0	0	17.4	21.6
Aluminum Foil (0.0025 in.)	28.3	17.6	24.8	20.3	41.3	43.7	23.0	22.8
Adhesion Properties:								
SAFT (C)	90	87	82	70	95	95	83	88
Pop-Open (C)	80	28	25	42	100	97	25	25
Low Temp. Bond Strength (–34 C)	PT	PT	PT	PT	PT	PT	PT	PT
Cold Flexibility (C)	25	25	25	25	>–34	>–34	25	25

	No. 9	No. 10	No. 11	No. 12	No. 13	No. 14	No. 15	No. 16
				(Polyethylene/Resin)				
Petrothene NA596	40	40	40	40	40	40	40	40
Wingtack® 95	40	40	40	40	40	40	40	40
Microwax (m.p. 185 F)	20	20	20	20	—	—	15	15
BHT Antioxidant	1	1	1	1	1	1	1	1
L-14 Polybutene	—	15	—	—	—	—	25	25
H-100 Polybutene	—	—	30	—	—	—	—	—
H-300 Polybutene	—	—	—	40	—	—	—	—
H-1500 Polybutene	—	—	—	—	20	—	5	—
H-1900 Polybutene	—	—	—	—	—	20	—	5

	No. 9	No. 10	No. 11	No. 12	No. 13	No. 14	No. 15	No. 16
Properties:								
Visc., cps @ 300 F	3,960	1,820	1,550	1,550	10,840	11,500	1,575	1,560
Softening Point, F	208	204	203	206	213	215	204	205
Tensile Strength, psi	80.3	241.4	205.0	188.5	257	258	68	73
Elongation, %	0	44	44	50	202	205	14	12
Shore A Hardness	87	65	58	54	63	62	36	37
Needle Penetration	5	18	22	30	12	11	26	30
Test Results (180° Peel Adhesion, oz/in. width)								
Natural Kraft Paper (60 lb/ream)	19.1	24.8	28.6	48.6	41.6	38.4	9.1*	17.5
Polyethylene Film (untreated)	15.7*	18.6	21.3	21.2	0	3.6	21.4	23.8
Aluminum Foil (0.0025 in.)	27.2	24.8	22.1	26.4	44.3	44.3	18.1	21.0

* Denotes no substrate failure

	No. 9	No. 10	No. 11	No. 12	No. 13	No. 14	No. 15	No. 16
Adhesion Properties:								
SAFT (C)	80	75	57	57	83	77	68	63
Pop-Open (C)	92	42	60	52	97	95	25	25
Low Temp. Bond Strength (−34 C)	PT	PT	PT	PT	PT	PT	PT	PT
Cold Flexibility (C)	25	25	25	25	>−34	>−34	25	25

	No. 17	No. 18	No. 19	No. 20	No. 21	No. 22
			(Polyethylene/Resin)			
Petrothene NA601	40	40	40	40	40	40
Wingtack® 95	40	40	40	40	40	40
Microwax (m.p. 185 F)	20	20	20	20	—	—
BHT Antioxidant	1	1	1	1	1	1
L-14 Polybutene	—	5	—	—	—	—
H-100 Polybutene	—	—	10	—	—	—
H-300 Polybutene	—	—	—	20	—	—
H-1500 Polybutene	—	—	—	—	20	—
H-1900 Polybutene	—	—	—	—	—	20
Properties:						
Visc, cps @ 300 F	660	500	470	475	1,313	1,375
Softening Point, F	200	198	199	198	204	203
Tensile Strength, psi	638.4	380.4	289.6	232.8	185	176
Elongation, %	90	53	58	44	124	100
Shore A Hardness	86	78	67	58	55	57
Needle Penetration	5	7	9	13	14	14

	No. 17	No. 18	No. 19	No. 20	No. 21	No. 22
Test Results (180° Peel Adhesion, oz/in. width)						
Natural Kraft Paper (60 lb/ream)	23.0	20.1*	47.2	52.0	32.0	24.0
Polyethylene Film (untreated)	10.2*	24.6	24.7	21.3	15.0*	8.5*
Aluminum Foil (0.0025 in.)	26.2*	28.3	29.3	30.1	40.0	39.5
Adhesion Properties:						
SAFT (C)	65	57	43	40	60	60
Pop-Open (C)	67	45	30	30	65	75
Low Temp. Bond Strength (–34 C)	NT	NT	NT	NT	NT	NT
Cold Flexibility (C)	25	25	25	25	25	25

* Denotes no substrate failure

No. 23

(Rosin/Resin)

DYLT	40.0
Dymerex	25.0
Piccopale® 110-SF	10.0
Atlantic 1116	25.0
BHT	0.5

Note:

This is a low cost hot-melt packaging adhesive.

Packaging Adhesive

(Water-Dispersed, Rubber/Resin)

	Formula No. 1	No. 2	No. 3	No. 4
Base Emulsion:				
Amoco® Resin 18-210	40	40	40	40
Amoco Polybutene H-100	40	40	40	40
Kraton® 1107	100	100	100	100
Xylene	100	100	100	100
Emulsifiers*	8	8	8	8
Irganox® 1010	2	2	2	2
Water	90	90	90	90
Tackifier:				
Wingtack® 76/Hexane				
Sol'n. (3:1)	380	380	380	380
Latex:				
Ucar® 154	0	63	126	252
Yield	760	823	886	1012

---

*   2 parts Triton® X-100; 2 parts Brij® 98 and 4 parts Tween® 80.

Procedure:

Heat **Resin 18** in solvent at 80 C until dissolved.   Add preheated (80 C) polybutene and stir. Pour solution over TPE and heat 15 min. in oven at 99 C.   Place on pebble mill for 2–3 h.   Alternately heat and roll until homogeneous.   Heat to 80 C and mix in high speed stirrer at 2000 rpm.

Gradually add preheated (80 C) emulsifiers and antioxidant. Add boiling water slowly until inversion occurs. Add remaining water. Mix for 10 min. Blend in tackifier and latex.

Properties:

Rolling Ball Tack (PSTC-6),				
cm	27.9	7.8	> 30.5	> 30.5
Shear Adhesion, s	9.0	24.8	481.1	694.8
Peel Adhesion, 180° (PSTC-1)				
N/mm width	0.4	1.5	1.6	1.3
lb/in.	2.4	8.3	8.9	7.4
Quick Stick Adhesion (PSTC-5)				
N/mm width	0.2	1.3	1.3	0.9
lb/in.	1.0	7.3	7.4	5.1

Bookbinding Adhesive

(Acrylic)

Gelva TS-70	90
Dibutyl Phthalate	6
Santolite MS-80	4

Hot-Melt Bookbinding Adhesive

Formula No. 1

(Ethylene Vinyl Acetate/Resin)

Zonarez B-100	30.0
Acintol® R Type 3A	20.0
EVA (28% VA-6 MI)	45.0
Microcrystalline Wax (m.p. 145 F)	5.0

Properties:

Sealing Temp.	350 F
Visc. @ 340 F	2930 cps
@ 350 F	2640 cps
Adhesion (Kraft-to-Kraft)	32 oz/in.

Tensile Properties:
    Tensile Strength   560 psi
    Elongation at Break   950%
    Break Energy   5875 in. lb/in.^{2}

This adhesive offers excellent strength and moderate viscosity at coating temperatures, exhibits excellent adhesion to coated stocks, and resists cold cracking.

No. 2

(Ethylene Vinyl Acetate/Resin)

Zonester® 85	30.0–40.0
EVA resin (28% VA-6 MI)	30.0–40.0
Paraffin Wax ($\approx$ m.p. 150 F)	20.0–30.0
Antioxidant	0.1–0.5

No. 3

(Ethylene Vinyl Acetate/Resin)

Elvax 260	32.9
Super Sta-Tac® 80	11.0
Super Sta-Tac® 100	28.0
Paraffin Wax (m.p. 150 F)	28.0
Antioxidant	0.1

Properties:
Visc. @ 300 F	3700–3800 cps
@ 350 F	1600–1700 cps
Elongation ($^1/_4$ in. × 20 mil film pulled at 2 in./min.)	275–300%

No. 4

(Ethylene Vinyl Acetate/Resin)

Elvax 420	54.0
Elvax 250	6.0
Nirez® V-2040	40.0
Wax (m.p. 150 F)	12.0
Antioxidant	1.0

Properties:

Visc. @ 300 F	25,000–26,000 cps
@ 350 F	11,000–12,000 cps

Elongation ($^1/_4$ in. × 20 mil film pulled at 2 in./min.) 400+%

Aging Characteristics Good color and no char or
(after 100 h @ 350 F) surface skinning

No. 5

(Ethylene Vinyl Acetate/Resin)

EVA-607	45
Staybelite®	20.0
Pentalyn A	30.0
Multiwax 180-M	5.0
BHT	0.5

No. 6

(Ethylene Vinyl Acetate/Resin)

EVA	25–50
Tackifying Resin	30–60
Wax	0–35
Filler	0–15
Antioxidant	0–1
Other Polymer	0–20

Note:

A typical recipe most popularly used is EVA (40 parts) + tackifying resin (40 parts) + petroleum wax (20 parts) + antioxidant (0.2 parts).

No. 7

(Ethylene Vinyl Acetate/Resin)

EVA	25–40.0
Tackifying Resin	30–60.0
Wax	10–35.0
Filler	0–10.0
Antioxidant	0–0.5
APP	0–50.0

For tackifying resins, gum rosin WW and petroleum resins are used in combination most popularly. As for the wax, paraffin wax with melting point from 130–150 is used. For fillers, small quantities of $CaCO_3$ or clay are sometimes used. BHT is typical as antioxidant. APP is used in most cases for reducing the cost.

No. 8

(Rubber/Resin)

Kraton® D 1102	100
Foral® 85	100
Picco® 6140	100
Paraffin Wax (m.p. 60 C)	100
Antioxidant	3

Hot-Melt Paperback Bookbinding Adhesive

	Formula No. 1	No. 2
	(Ethylene Vinyl Acetate/Rosin)	
Wingtack® 95	—	40
Micro Wax (m.p. 180 F)	39	25
Elvax 260	30	34
Rosin Ester	30	—
WingStay S	—	—

Hot-Melt Unstapled Bookbinding Adhesive

(Ethylene Vinyl Acetate/Resin)

EVA	30–50
Tackifying Resin	30–60
Wax	0–20
Titanium Oxide	0–15
Antioxidant	0.1–1
Synthetic Rubber	0–10

Note:

For EVA, any one of or combinations of **Nipoflex 710, 750,** and **720** are used. The tackifying resin is selected from among rosin polyol ester, its hydrogenated one, low molecular weight styrene resins, and others. The wax used most popularly are microcrystalline wax with melting point of about 175 F or the same in combination with paraffin wax with melting point of about 145 F. When higher heat resistance is required, a small quantity of coal wax is sometimes added. In some cases, wax and titanium oxide are not used at all. Typical antioxidants used for this purpose are BHT or 4,4′-methylene bis (2,6-di-*tert*-butyl phenol). When cold resistance is particularly required, a small quantity of butyl rubber is added.

Bag-Seam Adhesive

No. 1

(Ethylene Vinyl Acetate)

Airflex 300	90.00
Benzoflex 9-88	4.70
Polyox 750 WSN	0.10
Surfynol® D-201	0.20
Water	5.00

No. 2

(Water-Based/Dextrin)

Water	66.9
Stadex 27	27.9
Borax (5 mole)	4.2
Defoamer	0.1

Heat to 195 F. Hold 15 min. Cool below 150 F. Add:

Sodium Hydroxide (50% Sol'n.)	0.8
Preservative	0.1

Mix well; dilute, if required, to adjust viscosity.

Properties:

Solids (Refractometer)	28–30%
pH	9.2–9.4
Visc. (Brookfield) @ 80 F	1500–2000 cps

No. 3

(Water-Based/Starch)

Water	67.5
Obax S	30.6
Urea-Formaldehyde Resin (85% solids)	1.9

Procedure:
Cook at 190–195 F for 15 min. Cool.

Properties:
Solids (Refractometer)	25–27%
pH	≈ 5.5
Visc. (Brookfield) @ 80 F	1000–1500 cps

Grocery-Bag-Seam Paste

Formula No. 1

(Water-Based/Starch)

Water	72.8
Staclipse L	23.7
Borax (10 mole)	2.9
Preservative	0.2

Cook 15 min. @ 190–195 F. Cool to 140 F. Add:

Caustic Soda (50% sol'n.)	0.4

Mix. Cool. Dilute if necessary.

Properties:
Solids (Refractometer)	24–25%
pH	9.2–9.5
Visc. (Brookfield) @ 80 F	≈ 2000 cps

No. 2

(Water-Based/Starch)

Water	75
Staybind A	25

Procedure:
Cook 15 min. @ 190–195 F. Cool. Dilute if necessary.

Properties:
Solids (Refractometer)	22–23%
pH	≈ 9.5
Visc. (Brookfield) @ 80 F	≈ 2000 cps

No. 3

(Water-Based/Starch)

Water	83.67
Staley Pearl Starch	9.00
Polyvinyl Alcohol (fully hydrolyzed)	6.00
Soap	0.10
Preservative	0.10
Urea-Formaldehyde Resin (65% solids)	1.00

Mix the above ingredients well at room temperature. Heat to 190–195 F and hold 15 min. Cool to 150 F and add:

20% Alum Sol'n.	0.13

No. 4

(Water-Based/Starch)

Water	56.75
Eclipse F	5.70
Polyvinyl Alcohol (fully hydrolyzed)	5.70
Clay	1.50
Defoamer	0.30
Soap	0.15
Preservative	0.15
Urea-Formaldehyde Resin (65% solids)	0.45

Mix until thoroughly dispersed. Heat to 195 F and hold for 20 min. Cool to 150 F and add:

Polyvinyl Acetate Emulsion (55% solids)	29.00
Dibutyl Phthalate	0.30
Phosphoric Acid	to adjust pH to 3.5

Sugar-Bag Bottom Paste

(Water-Based/Dextrin)

Water	53.00
Koldex 60	36.50
Borax (5 mole)	3.55
1300 Corn Syrup	6.62

Cook 15 min. @ 190–195 F. Cool to 150 F. Add:

Formaldehyde	0.33

Mix. Transfer to drums. Age 3–5 days before shipment.

Layflat Bottom Paste

(Water-Based/Starch)

Water	54.5
Obax B	33.0
Sodium Nitrate	7.5
Urea	5.0

Procedure:
Cook 15 min @ 190–195 F. Transfer to drums while hot.

Bottom Paste

(Water-Based/Starch)

Water	83.44
Staley Pearl Starch	11.00
Polyvinyl Alcohol (fully hydrolyzed)	4.00
Soap	0.13
Preservative	0.10
Urea-Formaldehyde Resin (65% solids)	1.00

Plasticized Nonwater-Resistant
Small-Bag Bottom Paste

Formula No. 1

(Water-Based/Starch)

Water	67.8
Soap	0.3
Urea	2.9
Obax B	29.0

Procedure:
 Cook 15 min. @ 175–180 F. Cool, dilute if required.

Properties:
 Solids (Refractometer)   27–28%
 Visc. (Brookfield) @ 80 F   60,000–70,000 cps

No. 2

(Water-Based/Starch)

Water	57.50
Soap	0.25
Sodium Nitrate	5.35
Obax S	24.60
Eclipse F	12.30

Procedure:
 Cook 15 min. @ 185–190 F. Cool, dilute as required.

Properties:
 Solids (Refractometer)   36–38%
 Visc. (Brookfield) @ 80 F   40,000–50,000 cps

No. 3

(Water-Based/Starch)

Water	64.8
Soap	0.5
Preservative	0.2
Sodium Nitrate	4.5
Eclipse G	30.0

Procedure:
Cook at 185–190 F. Cool, dilute as required.

Properties:
Solids (Refractometer)	29–30%
Visc. (Brookfield) @ 80 F	60,000–70,000 cps

Multiwall W&H Bottom Adhesive

(Water-Based/Starch)

Water	72.0
Obax B	13.2
Obax S	13.2
Urea-Formaldehyde Resin (65% solids)	1.6

Procedure:
Cook at 190–195 F for 15 min. Cool.

Properties:
Solids (Refractometer)	21–23%
pH	≈ 5.5
Visc. (Brookfield) @ 80 F	2500–3000 cps

Water-Resistant, Multiwall
St. Regis-Type Bottom Paste

(Water-Based/Starch)

Water	76.1
Obax B	23.5
Urea-Formaldehyde	1.4

Procedure:
Cook at 190–195 F for 15 min. Cool.

Properties:
Solids (Refractometer)	19–21%
pH	≈ 5.5
Visc. (Brookfield) @ 80 F	15,000–25,000 cps

Self-Seal Envelope Adhesive

Formula No. 1

(Natural Rubber Latex)

Natural Rubber Latex (ammonia preserved—60%)	100.0
Potassium Hydroxide Sol'n. (10%)	0.2
Zinc Diethyl Dithiocarbamate (50% aq. disp.)	0.5
Water	as required

No. 2

(Natural Rubber Latex)

Natural Rubber Latex (60%)	167.0
Potassium Hydroxide Sol'n. (10%)	2.0
Zinc Diethyl Dithiocarbamate (50% disp.)	1.0

Front-Seal Envelope Adhesive

(Water-Based/Dextrin)

Water	32.9
Suitable Dextrin*	63.5
Tributyl Phosphate	0.1
Sodium Bisulfite	0.1
Polyethylene Glycol 400	0.2

Heat to 195 F. Hold 1 h. Cool to 140 F. Add polyvinyl acetate emulsion.

Corn Syrup (or other suitable plasticizer)	1.1

Mix and cool.

Properties:

Solids (Refractometer)	64–67%
Visc. (Brookfield) @ 80 F	5,000–10,000 cps

---

* Potato, tapioca, or waxy maize-based canary dextrins give best properties. **Staley's 955 SR Tapioca** dextrin, or **Stadex 230** dextrin should perform well.

Back-Gum Envelope Adhesives

Formula No. 1

(Water-Based/Dextrin)

Water	29.5
Stadex 90	32.8
Stadex 9	14.2
Urea	20.0
Monosodium Phosphate	2.9
Sodium Bisulfite	0.2
Preservative	0.2
Defoamer	0.2

Procedure:
Heat to 170 F. Hold 10 min. Cool.

Properties:
Visc. (Brookfield) @ 80 F   ≈ 20,000 cps

No. 2

(Water-Based/Dextrin)

#11 Tapioca Dextrin	48.9
Dextrose	48.9
Monosodium Phosphate	1.5
Sodium Bisulfite	0.2
Preservative	0.3
Defoamer	0.2
Water	to attain desired visc. (≈ 50% solids)

Procedure:
Cook to 190 F. Hold 10 min. Cool.

Hot-Melt Adhesive for Soap Wrappers and Bands

Formula No. 1

(Ethylene Vinyl Acetate/Resin)

Ultraflex®	50
Elvax 250	30
Paraffin Wax	20

Properties:

Visc.—Initial, ASTM D-3236 @ 250 F (121 C)   2450 cps
                        @ 275 F (135 C)   1650 cps
                        @300 F (149 C)   1200 cps
                        @ 325 F (163 C)   850 cps
                        @350 F (177 C)   680 cps

Visc.—Aged, after 32 h @ 300 F (149 C), ASTM D-3236
                        @ 300 F (149 C)   1175 cps

Aged Appearance—
  After 72 h @ 300 F (149 C)   No color change
Penetration, ASTM D-1321 @ 100 F (38 C)   18
Softening Point, ASTM E-28   157 F (69 C)

No. 2

(Ethylene Vinyl Acetate/Resin)

Ultraflex®	60
Elvax 250	35
Nirez® 1085	5

Properties:

Visc.—Initial, ASTM D-3236 @ 250 F (121 C)   7200 cps
                        @ 275 F (135 C)   4650 cps
                        @ 300 F (149 C)   3150 cps
                        @ 325 F (163 C)   2200 cps
                        @ 350 F (177 C)   1700 cps

Visc.—Aged, after 32 h @ 300 F (149 C), ASTM D-3236
                        @ 300 F (149 C)   3200 cps

Aged Appearance—
  After 72 h @ 300 F (149 C)   No color change
Penetration, ASTM D-1321 @ 100 F (38 C)   21.5
Softening Point, ASTM E-28   160 F (71 C)

Note:
  Good low-temperature adhesion.

No. 3

(Ethylene Vinyl Acetate/Resin)

Ultraflex®	65
Elvax 250	35

Properties:

Visc.—Initial, ASTM D-3236	@ 250 F (121 C)	6350 cps
	@ 275 F (135 C)	4200 cps
	@300 F (149 C)	2900 cps
	@ 325 F (163 C)	2100 cps
	@350 F (177 C)	1500 cps
Visc.—Aged, after 32 h @ 300 F (149 C), ASTM D-3236		
	@ 300 F (149 C)	2800 cps
Aged Appearance—		
After 72 h @ 300 F (149 C)		No color change
Penetration, ASTM D-1321 @ 100 F (38 C)		18
Softening Point, ASTM E-28		162 F (72 C)

School Paste

Formula No. 1

(Water-Based/Dextrin)

Water	62.6
Stadex Dextrin No. 9	18.0
Koldex Dextrin No. 60	14.0
Benzoic Acid	0.3
Oil of Sassafras or Methyl Salicylate	0.1
Corn Syrup (43 Bé)	5.0

Procedure:

Heat to 200 F. Hold at 200 F for 30 min. Cool to 140 F. Bottle or transfer to containers.

Properties:

Solids (Refractometer)	28–30%
pH	≈ 4
Fresh Visc. @ 80 F	Thin, fluid
Consistency, aged 3–4 days	Firm paste

No. 2

(Water-Based/Dextrin)

Water	59.6
Koldex Dextrin No. 60	40.0
Benzoic Acid	0.3
Oil of Sassafras or Methyl Salicylate	0.1

Procedure:
Heat to 200 F. Hold at 200 F for 30 min. Cool to 140 F. Bottle.

Properties:

Solids (Refractometer)	38–39%
pH	≈ 4
Consistency, aged 3–5 days	Firm paste

Tube Glue for Paper

(Water-Based)

Water (@ 20±5 C)	85.6
Check-R-Lamn Powder	12.0
Sodium Orthosilicate (anhyd.)	1.2
Boric Acid	0.6
Water	0.6

Procedure:
A brief wet-in period of the dry powder in ambient water (3–5 min) in a well-agitated, low-shear mixing tank should occur. An alkaline dispersing agent is then added and mixed for 5–10 min. Then a buffering agent is mixed in (5–10 min) and the glue is ready to use.

Properties:

Solids	13.5%
pH	10.8

Solid Paper-Fiber Adhesive

(Water-Based)

Penford Gum 280 (or **380**)	70.0
Coating Clay	30.0
Caustic Soda	0.25–0.5
Water	186.0

Procedure:

All ingredients are heated to 190–195 F with agitation and then cooled to the final operating temperature of 130–150 F.

Note:

This formula gives an adhesive the viscosity of which is remarkably stable on aging.

The addition of 10% urea-formaldehyde resin solids (on total solids) to the above adhesive prepared without the caustic soda gives an adhesive suitable for use in waterproof board which will withstand prolonged soaking in water without delamination. Such a mixture must, of course, be used with an appropriate catalyst such as one of the ammonium salts.

Paper-Core Adhesive

(Polybutene)

Polybutene 128	85
NI-W	15

Paper-to-Aluminum-Foil Adhesive

(Rubber/Resin)

Hycar® 1562X103	253.0
Vinsol®	75.0
Acrysol® GS	8.3

Note:

Point of laminate failure: Paper fibers.

Initial compound visc.   740 cP

Paper-to-Vinyl Film Adhesive

	Formula No. 1	No. 2	No. 3
		(Dry/Latex)	
Geon® 460X1 (51%)	100.00	100.0	—
Geon® 450X20 (55%)	—	—	100.0
Tetrasodium Pyrophosphate (100%)	0.10	—	—
Carbopol® 934 (5%)	1.26	0.5	—
Methocel® HG-65 or			
Methocel® MC 4000 (5%)	—	—	0.8
Properties:			
pH	7.2	5.5	—
Visc., cP	28,000	2900	3000

Note:
Ammonium hydroxide used to adjust pH for desired viscosity.

Corrugated-Paper Adhesive

(Water-Based)

Hot Set:	
Water	240
Purina Protein	50
Pearl Starch	10
Sodium Silicate	55
Properties:	
pH	10.5

Procedure:
Flush mixing tank clean with water and drain empty. Any residual silicate will coat the protein with a gelatinous film which cannot be readily dispersed and must be discarded. Add 240 lb (28.8 gal) water to mixing drum ($22^1/_4$ in. dia.). With agitator running, add 50 lb (1 bag) **Purina Protein** slowly to prevent lumping and continue mixing. Add 10 lb pearl starch slowly, sifting to prevent lumping and mixing until the entire mixture is smooth. Add 55 lb (4.74 gal) sodium silicate very rapidly anticipating thickening of the mix when only part of the silicate is

present, and thinning rapidly as the full amount is dispersed. Continue mixing for at least 15 min to break down this dispersion which stabilizes at a relatively thin viscosity.

Cold Set:

Water	250.000
Sodium Silicate	3.000
Kaolin Clay	100.000
Sodium Sulfite	0.125
Purina Protein	50.000
Sodium Silicate	48.000

Properties:

pH	10.5

Procedure:

Flush mixing drum clean with water and drain empty. Any residual silicate will coat the protein with a gelatinous film which cannot be dispersed and must be discarded due to lumping. Add 18 in. (30 gal) water to mixing drum (22$^1/_4$ in. dia.). Add 1 qt sodium silicate and 250 ppm disodium ethylene diamine tetraacetate (EDTA). Mix sufficiently to disperse. Add 100 lb (2 bags) kaolin clay slowly to prevent lumping, agitating continuously for 5 min. Add 1 oz sodium sulfite while continuing agitation. Add 50 lb (1 bag) **Purina Protein** slowly sifting to prevent lumping and continue mixing 5–15 min until entire mixture is smooth. Add 48 lb (4 gal) sodium silicate very rapidly, anticipating thickening of the mix when only part of the silicate is present and thinning rapidly as the full amount is dispersed. Continue mixing for about 15 min to break down the dispersion which stabilizes at a relatively fluid viscosity.

Paper-to-Paper Adhesive

	Formula No. 1		No. 2		No. 3	
			(Latex/Resin)			
	(Dry)	*(Wet)*	*(Dry)*	*(Wet)*	*(Dry)*	*(Wet)*
Good-Rite® 2570X5	40	84.0	40	84.0	—	—
Good-Rite® 2570X54	—	—	—	—	40	100.0
Vinsol® Emulsion	25	62.5	25	62.5	25	62.5
Dresinol 42	—	—	25	62.5	25	62.5
Dresinol 205	25	55.8	—	—	—	—
Hydrasperse	10	15.4	—	—	—	—
Dixie Clay	—	—	10	15.4	10	15.4

No. 4

(Latex/Resin)

Hycar® 1562X103	40.00
Vinsol® Emulsion	40.00
Good-Rite® K-718	20.00

No. 5

(Latex/Resin)

Good-Rite® 2570X22	40
Vinsol® Emulsion	25
Dresinol 42	25
Hydrasperse	10

Paper-to-Burlap Adhesive

	Formula No. 1		No. 2	
		Latex		
	(Dry)	*(Wet)*	*(Dry)*	*(Wet)*
Hycar® 2570X1 (56%)	100	180	100	180
Tamol® 731 (25%)	2	8	2	8
No. 10 Whiting	100	100	150	150
Alcogum AN 10 (5% sol'n.)	—	—	2	40
Water	—	—	—	150

Paper-Match Adhesive

(Starch)

Waxy-Maize Starch	77.5
Humectant	10.0
Polyvinyl Alcohol	0.5
Dextrin	4.0
Urea	8.0

Release Coating for Paper

(Water-Based/Silicone)

Water	79.5
CMC (solids)	5.0
Glacial Acetic Acid	0.5
Syl-off 22	12.5
Dow Corning® 22A	2.5

Note:

Glacial acetic acid is added to the bath to prolong the effective life of the catalyzed emulsion. Without acetic acid, pot life is from 1–2 h at 75 F (24 C). Acidified, the pot life of the coating bath is ≈ 6 h at 75 F (24 C). Pot life is lengthened by cooling. For maximum life, the coating bath should be kept at normal room temperature or below.

Layflat "Mounting" Adhesive

(Water-Based/Dextrin)

Water	32.5
Staclipse I	42.4
Sodium Nitrate	15.0
Calcium Chloride	5.0
Ethylene Glycol	5.0
Dowicide G or other suitable preservative	0.1

Procedure:

Mix at room temperature. Cook 15 min @ 180–185 F. Cool.

Properties:

Solids (Refractometer)	60–62%
Visc.—Initial (Brookfield) @ 80 F	3000–4000 cps

Collating Adhesive

(Water-Based/Dextrin)

Stadex Dextrin 126	150
Urea	100
Pluronic L-61 or other suitable defoamer	1
Dowicide A or other suitable preservative	1
Water	110

Cook to 175 F for 30 min. Add:

Borax (10 mol)	15

Agitate until borax is dissolved. Cool to 120 F.

Properties:

Solids	66.0%
pH	8.70
Visc. @ 80 F	5000–7000 cps

Waterproof Casein Adhesive

(Lactic Acid)

Dissolve:

Casein	100
Urea	80

in

Water	100

Add:

Sodium Fluoride	5
Borax	10

Note:

Viscosity can be regulated by pH (use diluted HCl). This adhesive is used in beer bottle labels.

Glue-Lap Adhesive

(Ethylene Vinyl Acetate)

Airflex 300	86.70
Benzoflex 9-88	7.00
ASP® 600	2.00
Surfynol® D-201	0.20
Water	4.10

Remoistening Adhesive

(Water-Based/Dextrin)

Yellow Potato Dextrin	100	lb
Water	56.0	lb
Borax	3.0	lb
Sodium Perborate	1.0	lb
Glyceryl Borate	9.0	lb
Formalin	10.0	fl oz
Santobrite	0.2	lb

Nonwoven Web Binder

	Formula No. 1	No. 2	No. 3	No. 4	No. 5
			(Acrylic Latex)		
Hycar® 2600X120	60.0	—	—	—	—
Hycar® 2679	—	60.0	—	—	—
Hycar® 2600X104	—	—	60.0	60.0	60.0
Permafresh LF	—	—	—	3.0	6.0
Ammonium Chloride	—	—	—	0.3	0.6
Oxalic Acid	0.7	0.7	0.7	0.7	0.8

Procedure:

After saturation, samples are nipped at 20 lb of pressure and immediately dried for 5 min at 135 C (275 F).

Hot-Melt, High-Gloss Display Coatings

Formula No. 1

(Ethylene Vinyl Acetate/Resin)

Zonarez 7115	15.0
EVA (18% VA-150 MI)	20.0
Epolene C-13	8.0
Paraffin Wax (m.p. 150 F)	52.0
Microcrystalline Wax (m.p. 160 F)	5.0

Properties:

Visc. @ 250 F	885 cps
@ 350 F	240 cps
Gloss	68
MVTR	0.08 g H_2O/100 in.^{2}-24 h

Note:

This is a medium-viscosity, high-quality coating with good scuff and mar resistance. It exhibits gloss and gloss stability.

No. 2

(Low-Viscosity, Polyethylene/Resin)

Zonarez 7115	5.0
Epolene C-10	20.0
Paraffin Wax (m.p. 145 F)	45.0
Microcrystalline Wax (m.p. 165 F)	30.0

Properties:

Visc.@ 250 F	31 cps
Gloss	60

Note:

This coating is formulated to give a low-viscosity, low-coast coating which also has good gloss. This coating sacrifices some of the gloss of a high-quality coating for the advantage of low cost.

Hot-Melt, General-Barrier Coating for Corrugated Containers

(Ethyl Vinyl Acetate/Resin)

Zonarez 7125	15.0
EVA (28% VA-6 MI)	25.0
Epolene C-16	10.0
Paraffin Wax (m.p. 150 F)	42.0
Microcrystalline Wax (m.p 180 F)	8.0

Properties:
Visc. @ 250 F	1375 cps
@ 350 F	493 cps
Gloss	50
MVTR	0.09 g H_2O/100 in.^{2}-24 h
Adhesion (Kraft to Kraft)	20 oz/in.

Note:

This coating formulation is specifically designed for application by curtain coating. It is best suited for containers of fruit, vegetables, and meat. This formulation provides both high-barrier properties and decorative appeal.

Hot-Melt, Form-and-Fill Package Coating

(Ethylene Vinyl Acetate/Resin)

Ultrathene UE 634-04 (28% VA, 6 MI)	20
Styrene Copolymer Resin (high soft. pt.)	20
Paraffin Wax (m.p. 150 F)	60

Note:

This is a starting formulation providing good hot tack but some sacrifice in gloss.

Chapter II

CONSTRUCTION ADHESIVES

Construction Adhesives

Formula No. 1

(Rubber/Resin)

Kraton® D 1101	100
Cumar® LX-509	75
Atomite®	350
Stabilizers:	
Zinc Dibutyl Dithiocarbamate	2
Plastanox® 2246	1
Epon® 1002	2
Rule 66 Solvent	420

Properties:	
Solid Concentration	69%
Lap Shear Test	
Strength at 77 F	510 psi
Strength at 180 F	290 psi
Upper temperature limit	294 F
Peel Adhesion	
Canvas/Aluminum	12 pli
Canvas/Mild Steel	15 pli
Canvas/Plywood	16 pli
Tensile Strength (T_B)	1220 psi
Elongation at Break (E_B)	16%
Hardness (Shore D)	58
Flexibility (180° bend)	Good

	No. 2	No. 3	No. 4	No. 5	No. 6	No. 7
			(Rubber/Resin)			
Kraton® D 1101	100	100	100	100	100.0	100
Cumar® LX-509	75	75	75	50	37.5	25
Floral 105	—	25	50	50	37.5	50
Stabilizer:						
Zinc Dibutyl Thiocarbamate	2	2	2	2	2.0	2
Plastanox® 2246	1	1	1	1	1.0	1
Epon® 1002	2	2	2	2	2.0	2
Solvent			as required for useful viscosity			

	No. 2	No. 3	No. 4	No. 5 (cont'd.)	No. 6	No. 7
Properties:						
Total Solids	180	205	230	205	180	180
Yield Point, psi	1170	960	840	660	600	490
Tensile Strength (T_B), psi	4550	3890	3340	3670	3320	3580
Elongation (E_B), %	710	730	720	840	820	930
Shear Adhesion Failure Temp., °F	206	192	177	170	171	152
Peel Adhesion (180°), pli						
Canvas/Aluminum	5.7	6.2	8.5	5.7	4.0	24.5
Canvas/Mild Steel	11.7	13.5	15.0	10.0	6.7	24.5
Canvas/Plywood	20.0	23.0	32.0	21.0	20.0	35.0
Effect of oxygen bomb aging:						
% tensile properties retained after 1000 h						
T_B	98	99	99	90	108	71
E_B	99	101	106	108	112	104
% tensile properties retained after 2000 h						
T_B	86	77	46	42	58	38
E_B	104	107	119	129	130	131

	No. 8	No. 9	No. 10	No. 11
		(Rubber/Resin)		
Kraton® D 1101	100.0	100.0	100.0	100.0
Cumar® LX-509	37.5	37.5	37.5	37.5
Floral 85	37.5	—	—	—
Floral 105	—	37.5	—	—
Wingtack® 95	—	—	37.5	—
Wingtack® 115	—	—	—	37.5
Stabilizers:				
Zinc Dibutyl Thiocarbamate	2	2	2	2
Plastanox® 2246	1	1	1	1
Epon® 1002	2	2	2	2
Solvents		—— to give useful viscosity ——		

Properties:				
Total Solids	180	180	180	180
Yield Point, psi	620	600	670	770
Tensile Strength (T_B), psi	3550	3320	3500	3640
Elongation (E_B), %	850	820	850	830
Shear Adhesion Failure				
Temp., °F	164	171	180	185
Peel Adhesion, pli				
Canvas/Aluminum	3.2	4.0	2.0	4.3
Canvas/Mild Steel	7.5	6.7	3.5	5.2
Canvas/Plywood	24.0	20.0	32.0	15.0

	No. 12	No. 13	No. 14	No. 15	No. 16	No. 17
			(Rubber/Resin)			
Kraton® D 1101	100	100	100	100	100	100
Cumar® LX-509	75	75	75	75	75	75
Stabilizers:						
Zinc Dibutyl Thiocarbamate	2	2	2	2	2	2
Plastanox® 2246	1	1	1	1	1	1
Epon® 1002	2	2	2	2	2	2
Filler:						
Atomite®	—	200	350	500	—	—
Sierra Supreme 325	—	—	—	—	350	—
Clay (Type 50, soft)	—	—	—	—	—	350
Rule 66 Solvent	420	420	420	420	420	420

	No. 12	No. 13	No. 14	No. 15	No. 16	No. 17
			(cont'd.)			
Properties:						
Total Solids	180	380	530	680	530	530
Solids Concentration, %wt.	30	48	56	62	56	56
Lap Shear Tensile Strength						
at 77 F, psi	—	540	510	425	410	440
at 180 F, psi	—	270	290	290	220	320
Shear Adhesion Failure Temp., °F	206	287	294	283	278	309
Peel Adhesion, pli						
Canvas/Aluminum	5.7	8	12	10	7	10
Canvas/Mild Steel	11.7	18	15	12	7	11
Canvas/Plywood	20.0	21	16	11	9	10
Tensile Strength (T_B), psi	4550	1020	1220	1210	1260	1250
Elongation at Break (E_B), %	710	80	16	5	4	15
Hardness, Shore A	—	54	58	63	57	60
Flexibility (180° bend)	Good	Good	Good	Sl. cracks	Sl. cracks	Good

No. 18

(Rubber/Resin)

Kraton® D 1101	100
Cumar® LX-509	75
Atomite®	350
Butazate®	2
Epon® 1002	2
Plastanox® 2246	1
Super VM&P Naphtha	340
Toluene	80

	No. 19	No. 20	No. 21
	(Styrene-Butadiene/Resin)		
Solprene 411	100.0	100.0	100.0
Picco® 6100 1¹/₂	—	100.0	—
Picco® 6140 3	200.0	100.0	250.0
Hard Clay	200.0	200.0	250.0
Asbestos Fibers	12.5	12.5	12.5
Toluene	25.0	25.0	25.0
Textile Spirits	350.0	325.0	300.0
Ethyl Alcohol	12.5	12.5	12.5
AgeRite D	2.0	2.0	2.0
DLTDP	0.5	0.5	0.5

Properties:

Test Condition[1]	*Compression*	*Shear Strength*	*lb (kN)*
Dry Lumber	506 (2.25)	331 (1.47)	467 (2.08)
Gap Filling Test	392 (1.74)	199 (0.89)	467 (2.08)
Frozen Lumber	266 (1.18)	307·(1.37)	528 (2.35)
Wet Lumber[2]	206 (0.92)	322 (1.43)	189 (0.84)

---

[1] Similar to ARG-01 Specifications
[2] Tested in Tension

	No. 22	No. 23	No. 24	No. 25	No. 26
		(Rubber/Resin)			
A Hycar® ATBN					
(1300X16)	100	100	100	50	100
Thixcin R	10	10	10	10	10
Mistron Vapor®	30	—	—	—	—
#10 Whiting	—	200	200	130	200
Dioctyl Phthalate	40	20	20	25	20
B Epon® 828/					
BPA-100/24	—	20	—	—	—
Diethylene Glycol					
Diacrylate	10	—	10	—	—
Trimethylol Propane					
Triacrylate	—	—	—	—	5
Hycar® VTBN					
(1300X14)	—	—	—	50	—
Dioctyl Phthalate	—	20	20	25	20
Thixcin R	5	10	10	10	10
#10 Whiting	—	50	60	130	65
Total parts recipe					
% polymer in recipe	51	23	23	23	23
Properties:					
Tacky cure in, h	≈18	≈3	≈3	>3	≈3
Durometer A hardness:					
1 day	6/0	—	—	—	—
2 day	8/0	73/68	27/11	6/0	30/18
7 day	—	76/68	40/22	12/0	34/22
17 day	25/18	—	—	—	—

	No. 27	No. 28
	(Rubber/Resin)	
A Hycar® ATBN (1300X16)	100	100
Dioctyl Phthalate	100	100
B DGBEA	20	—
Trimethylol Propane Triacrylate	—	20

Properties:

Visc. (Brookfield) @ 27 C, cps after mixing	8500	8500
R.T. curing, 16 h	soft, tacky	soft, tacky

No. 29

(Rubber/Resin)

Ameripol 1013	60.0
Ameripol 1009	40.0
Super Sta-Tac® 80	75.0
Nirez® V-2040 HM	55.0
Cyanox 2246	130.0
Dixie Clay	130.0
Atomite®	130.0
Hexane	126.0
Toluene	85.0

Properties:

Total Solids	70%

No. 30

(SBR/Resin)

Ameripol 1013	60.0
Ameripol 1009	40.0
Super Sta-Tac® 100	70.0
Betaprene AC-130	40.0
Cyanox 2246	2.0
AF-950	130.0
ChemCarb® 11	130.0
Hexane	121.0
Toluene	81.0

Properties:

Total Solids	70%

No. 31

(SBR/Resin)

Ameripol 1013	60.0
Ameripol 1009	40.0
Synthe-Copal® 85	100.0
Beta-Tac® 160	70.0
Cyanox 2246	2.0
AF-950	270.0
ChemCarb® 11	265.0
Hexane	261.0
Toluene	174.0

Properties:
Total Solids   65%

No. 32

(Vinyl Acetate Ethylene/Resin)

Airflex 400 (56% T.S.)	545.0
Triton® X-405	4.5
Composition T	5.0
Ethylene Glycol	19.0
PMA-30	2.0
Amberol ST-140	71.0
Toluene	48.0
Atomite®	300.0
Carbopol® 934	5.0
Water	69.0
AMP-95	2.0

Procedure:

Add the ingredients in the order listed, being sure each is well mixed before the addition of the next. Dissolve the tackifier in toluene separately and add it to the major blend slowly to allow good dispersion for better emulsion stability. Disperse **Carbopol® 934** resin in water separately before adding to the other materials.

No. 33
(SBS Rubber)

Kraton® 1102	182.0
XL-30	45.0
Plastanox® 2243	2.4
Atomite®	200.0
Snobrite Clay	150.0
Toluene	174.0
Hexane	28.0
Acetone	176.0

Procedure:

Premix a 60% solids solution of **XL-30** in toluene. The **Kraton®** 1102 and **Plastanox®** 2243 are mixed into the **XL-30** toluene premix and the remaining solvent. By having the **XL-30** predissolved, the **Kraton®** 1102 will swell in a few hours with occasional stirring. About $^1/_3$ to $^1/_2$ of the resultant binder mass is charged into the mixer and pigmented. The amount of binder added should give an efficient stiff mix to insure quick wetting of the pigment. When a smooth, homogeneous mix is obtained ($\approx$ 30 min), the remainder of the binder mass is added and mixed until homogeneous.

Properties:

Density	9.6 lb/gal
Nonvolatile by volume	44.3%
Nonvolatile by weight	60.5%
Shear Strength Pine to Pine Bonding:	
ASTM D-906	546.6 lb/in.^{2}
Cure: 2 days @ 50 C, 4 days R.T.	
Dry Pine to Dry Pine	598.1 lb/in.^{2}
Cure: 28 days @ 100 F	
Dry Pine to Wet Pine	255.2 lb/in.^{2}
Cure: 28 days @ 100 F	
Dry Pine to Frozen Pine	436.2 lb/in.^{2}
Cure: 7 days @ 0 F + 21 days @ 40 F	
Dry Pine to Dry Pine	616.0 lb/in.^{2}
Cure: 500 h @ 150 F	
Durability:	
Moisture Resistance	521.9 lb/in.^{2}
Cure: 28 days @ R.T.	
AFG-01 Oxidation Resistance Test	Pass
(ASTM D572-67)	

No. 34

(Solvent-Based, Rubber/Resin)

Kraton® D-1101	100
Cumar® LX-509	75
Calcium Carbonate	350
ZDBT	2
Plastanox® 2246	1
Epon® 1002 F	2
Toluene	70
Naphtha	350

Construction Sealants

Formula No. 1

(For Glass Unit, Butyl Rubber)

A	**Butyl LM 430**	60.0
	Toluene	15.0
	Thixseal 436	6.0
B	Epoxy Silane	4.0
	Indopol H-1900	10.0
	Age Rite White	1.5
	p-Quinone Dioxime	3.5
	Omya® BLH	75.0
	Toluene	30.0

Procedure:

Thoroughly mix Part A at a minimum temperature of 158 F (70 C). Stir Part B into A, then mill paint for good dispersion.

No. 2

(Butyl Rubber)

Butyl LM 430	40.0
PbO_2	7.5
Cab-O-Sil® M5	5–10.0
Epon® 872	0–10.0
Toluene	20–40.0

Note:

Epon® 872 can slow bonding rate so its contribution to flow control should be balanced against the bonding rate equipment.

	No. 3	No. 4
		(Butyl Rubber)
Enjay Butyl LM 430	5	50.0
Calcene TM	30	—
HiSil 215	—	20.0
Titanium Dioxide	—	10.0
Stearic Acid	—	1.0
ZnO	—	5.0
Indopol H-100	60	—
Silane A-187	4	—
GMF	3	—
Cab-O-Sil® M5	—	5.0
PbO_2 (VFC, 0.33μ)	—	7.5

Properties:
Pot Life 2 h
Cures Overnight
Adheres to glass and aluminum in 1 day

Note:

Toluene will cause a chemical dew point in an insulated glass unit. This is avoided by using activated carbon, or large-pore size Molecular Sieve and/or silica gel in the spacer.

	No. 6	No. 7	No. 8	No. 9	No. 10
		(Polybutene/Butyl Rubber)			
Bucar 5214	100.00	100.00	—	—	—
Polysar XL-50	—	—	100.0	100.0	—
Aid-10	—	—	—	—	100.0
Amoco® H-100	143.00	—	200.0	—	267.0
Amoco® H-300	—	121.00	—	200.0	—
Keltrol®	12.10	22.00	50.0	50.0	50.0
Stearic Acid	2.20	2.20	—	—	—
Thixatrol GST	—	—	6.6	6.6	10.0
Calcium Carbonate	480.00	480.00	653.0	653.0	582.0
Talc	242.00	244.00	417.0	417.0	433.0
Titanium Dioxide	12.10	11.00	67.0	67.0	67.0
Mineral Spirits	110.00	132.00	166.0	166.0	150.0
Cobalt Drier (6%)	0.22	0.33	0.5	0.5	0.5
Antioxidant	—	—	2.0	2.0	2.0
Super Beckacite® 2000	—	—	5.0	5.0	5.0

Procedure:

15 min	A third of the polybutene and remainder of calcium carbonate is added.
25 min	Remaining polybutene and half the talc are added.
30 min	Remaining talc, TiO_2, and half the solvent are added and the mass mixed until homogeneous.
35 min	Mix cobalt drier and remaining mineral spirits and add the mixture incrementally over a 10-min period.
50 min	Mass should be homogeneous and ready for dumping. Compression ram used 0–35 min.

Mixing procedure for **Polysar XL-50** formulations:

Start	Steam on to approximately 250 F. Add thixotrope.
1 min	Add polybutene, antioxidant, and phenolic resin.
5 min	Steam off; incrementally add fillers and pigment.
15 min	Cold water on, add **Polysar XL-50**, lower ram.
30 min	Add copolymer incrementally.
50 min	Mix for 5 min.
55 min	Dump.

Self-Adhering Construction Sealant Compound

(Silicone)

RTV-31U	47.3
Burgess #30 Clay	30.4
Thixcin R	5.2
A-1110 Silane Adhesion Promoter	1.5
Odorless Mineral Spirits	15.1

Note:

This formulation is hand gunnable; shows no sag on a vertical surface; is fully cured as a bead in 3 days while retaining slight surface tack; and has excellent wet adhesion to glass, masonry, and metals.

Black Construction Sealant

	Formula No. 1	No. 2
	(Butyl Rubber)	
Enjay Butyl LM 430	90.0	10.0
Mt Black	150.0	—
Titanium Dioxide	—	2.0
Irganox® 1010	2.0	—
Silane A-187	6.0	—
Epon® 872	5.0	—
GMF	3.5	—
2-Pyrrolidone	2.0	—
MnO_2	—	10.0
Neodecanoic Acid	—	0.5
Toluene (dry)	80.0	4.5

Properties:

Work life < 1 h

Cure Time ≈ $1/2$ day

Note:

The high loading of black results in a thixotropic consistency—often desirable.

Solution-Applied Contact
Assembly Adhesive
(Rubber/Resin)

Kraton® 1101	100.0
Pentalyn® H	37.5
Picco® N-100	37.5
Plastanox® 2246	0.6
Hexane/Toluene/Acetone Ratio	60/20/20

Assembly Adhesive
Formula No. 1
(Rubber/Resin)

Kraton® 1101	100.0
Foral® 105	70.0
Soft clay	200.0
Antioxidant 330	0.5
DLTDP	0.5

Properties:

Films cast from toluene pass the tentative ASTM specification "Adhesives for Field-Gluing Plywood and Lumber Floors to Wood Framing." The test calls for aging 500 h at 300 psig and 158 F in pure oxygen, cooling, and bending around a $1/4$-in. mandrel without breaking.

Note:

Saturated resins such as **Foral® 105** are needed to pass the ASTM aging test. Unsaturated resins embrittle and fail.

Uses:

Construction adhesive for wood/wood joints.

No. 2
(Rubber/Resin)

Kraton® 1101	100.0
Piccolyte® A-125	100.0
Shellwax 300	165.0
Antioxidant 330	0.5
DLTDP	0.5

Properties:

Melt Visc.	@ 300 F	160,000 cps
	@ 325 F	75,000 cps
	@ 350 F	8,000 cps
Migration Pt.		135 F

Note:

This adhesive may be applied by die coating as well as by nozzle. A coating of 0.67 mils was applied on an Egan D8-Coater in laminating paper to foil.

Uses:

Melt-applied bonding adhesive with good high temperature performance for paper, wood, textile, leather, or metal products.

No. 3

(Rubber/Oil-Based)

Kraton® 1101	100.0
Asphalt	775.0
Shellflex® 371N	125.0
Antioxidant 2246	0.5
DLTDP	0.5
Toluene	285.0

Properties:

Solution visc.	@ 32 F	10,000 cps
	@ 50 F	4,200 cps
	@ 75 F	1,600 cps
	@ 100 F	700 cps
Slump [1]		0.25 in.
Flow [1]		0.25 in.
Softening Pt. (Ring & Ball)		398 F

180° Peel Strength (after curing 2 days @ 75 F, 50% R.H.)

Bonded to Self[2]	5.2 pli
Bonded to Plywood[3]	5.1 pli
Bonded to Steel[3]	5.7 pli
Bonded to Portland Cement[3]	6.5 pli
Bonded to Galvanized Iron[3]	4.2 pli

---

[1] ASTM D-1191-52T modified
[2] ASTM D-1876-61T 6in./min separation
[3] ASTM D-903-49

Note:

Full cure is essentially reached in two days. Solution viscosity drops rapidly with increasing toluene. The asphalt used had a 25 pen., 128 F ring and ball softening point, 4250 poise viscosity at 140 F.

Uses:

Flooring cements and improved performance of asphalt cements.

Hot-Melt Assembly Adhesives

Formula	No. 1	No. 2	No. 3	No. 4	No. 5	No. 6	No. 7
				(Ethylene Vinyl Acetate)			
Elvax 460	35	35	35	35	35	35	35
K-1120	30	20	10	0	20	10	0
Wax (m.p. 185 F)	35	35	35	35	35	35	35
Irganox® 1010	1	1	1	1	1	1	1
R-18-210	—	10	20	30	—	—	—
R-18-290	—	—	—	—	10	20	30
Properties:							
Peel Adhesion, oz/in. width							
Kraft/Kraft	21.28	19.41	17.92	21.44	20.32	17.49	14.83
Kraft/Foil	16.16	17.49	17.65	15.41	17.12	19.09	22.19
Kraft/PE	14.67	17.33	12.85	16.16	16.05	14.99	14.88
Shear Adhesion							
Fail Temp., °C	80	80	80	80	80	80	80
Low Temp. Bond							
Strength @ –32 C				Paper/fiber tear			

	No. 8	No. 9	No. 10	No. 11	No. 12	No. 13
			(Ethylene Vinyl Acetate)			
Elvax 46	30	25	20	30	25	20
K-1120	30	30	30	30	30	30
Wax (m.p. 185 F)	35	35	35	35	35	35
Irganox® 1010	1	1	1	1	1	1
R-18-210	5	10	15	—	—	—
R-18-290	—	—	—	5	10	15
Properties:						
Peel Adhesion, oz/in. width						
Kraft/Kraft	17.60	15.68	14.24	16.80	16.75	15.41
Kraft/Foil	15.36	13.44	12.85	15.89	13.55	12.75
Kraft/PE	14.88	12.43	8.85	13.49	12.80	11.20
Shear Adhesion						
Fail Temp., °C	80	80	80	75	80	80
Low Temp. Bond						
Strength @ –32 C	———— Paper/fiber tear ————					

	No. 14	No. 15	No. 16	No. 17	No. 18	No. 19	No. 20
			(Ethylene Vinyl Acetate)				
Elvax 420	35	35	35	35	35	35	35
K-1120	30	20	10	0	20	10	0
Wax (m.p. 185 F)	35	35	35	35	35	35	35
Irganox® 1010	1	1	1	1	1	1	1
R-18-210	—	10	20	30	—	20	—
R-18-290	—	—	—	—	10	—	30
Properties:							
Peel Adhesion, oz/in. width							
Kraft/Kraft	19.57	17.49	14.93	19.79	15.20	19.31	20.16
Kraft/Foil	17.49	16.85	14.72	18.13	14.03	16.53	14.56
Kraft/PE	15.20	13.17	14.13	16.32	16.80	11.84	16.11
Shear Adhesion							
Fail Temp., °C	70	70	70	75	75	75	70
Low Temp. Bond Strength @ –32 C			Paper/fiber tear				

	No. 21	No. 22	No. 23	No. 24	No. 25	No. 26
			(Ethylene Vinyl Acetate)			
Elvax 42	30	25	20	30	25	20
K-1120	30	30	30	30	30	30
Wax (m.p. 185 F)	35	35	35	35	35	35
Irganox® 1010	1	1	1	1	1	1
R-18-210	5	10	15	—	—	—
R-18-290	—	—	—	5	10	15
Properties:						
Peel Adhesion, oz/in. width						
Kraft/Kraft	18.45	16.00	16.75	18.13	14.19	7.89*
Kraft/Foil	13.76	14.77	16.48	17.81	18.56	6.72*
Kraft/PE	13.81	12.05	9.23	12.48	14.24	7.25*
Shear Adhesion Fail Temp., °C	70	75	85	75	75	75
Low Temp. Bond Strength @ –32 C			Paper/fiber tear			

* Denotes NO substrate failure

	No. 27	No. 28	No. 29	No. 30	No. 31	No. 32	No. 33
				(Ethylene Vinyl Acetate)			
Elvax 410	35	35	35	35	35	35	35
K-1120	30	20	10	0	20	10	0
Wax (m.p. 185 F)	35	35	35	35	35	35	35
Irganox® 1010	1	1	1	1	1	1	1
R-18-210	—	10	20	30	—	—	—
R-18-290	—	—	—	—	10	20	30
Properties:							
Peel Adhesion, oz/in. width							
Kraft/Kraft	17.97	18.08	18.77	18.99	21.39	19.63	22.93
Kraft/Foil	12.59	13.71	15.57	18.13	16.43	17.17	15.57
Kraft/PE	8.91*	5.71*	4.21*	0.0*	5.07*	5.76*	6.03*
Shear Adhesion							
Fail Temp., °C	70	75	60	55	70	70	65
Low Temp. Bond							
Strength @ −32 C				Paper/fiber tear			

* Denotes NO substrate failure

	No. 34	No. 35	No. 36	No. 37	No. 38	No. 39
			(Ethylene Vinyl Acetate)			
Elvax 410	30	25	20	30	25	20
K-1120	30	30	30	30	30	30
Wax (m.p. 185 F)	35	35	35	35	35	35
Irganox® 1010	1	1	1	1	1	1
R-18-210	5	10	15	—	—	—
R-18-290	—	—	—	5	10	15
Properties:						
Peel Adhesion, oz/in. width						
Kraft/Kraft	20.53	16.96	11.15*	15.84	18.67	13.81
Kraft/Foil	14.03	12.43	8.91*	14.19	16.48	13.44
Kraft/PE	5.60*	5.44*	2.61*	7.84*	4.56*	2.08*
Shear Adhesion Fail Temp., °C	70	70	70	70	65	70
Low Temp. Bond Strength @ –32 C	Paper/fiber tear					

* Denotes NO substrate failure

Wood and Plastic Assembly Adhesives

Formula No. 1

(Rubber/Resin)

Neoprene AF	100
Antioxidant	2
Zinc Oxide	5
Maglite D	8
t-Butyl Phenolic Resin	45
Toluene	213
Ethyl Acetate	213
Skellysolve B	213

Procedure:

Dissolve resin in 100 toluene then add remaining solvents. Break down **Neoprene AF** on rubber mill for 10 min. Add **Maglite D**, antioxidant, and zinc oxide. Add freshly milled compound to resin solution. Dissolve compound in the resin solution.

Properties:

Solids	20.2%

No. 2

(Rubber/Resin)

Neoprene AC (med. visc.)	100
Antioxidant	2
Maglite A	8
Zinc Oxide	5
t-Butyl Phenolic Resin	45
Toluene	166
Hexane	333
Methyl Ethyl Ketone	166

Procedure:

Dissolve resin in 100 toluene and add 4 **Maglite A**. Agitate until resinate formation is complete. Break down **Neoprene AC** on rubber mill for 10 min. Add **Maglite A**, antioxidant, and zinc oxide. Dissolve mill mix in mixture of remaining solvents. Add resinate solution and mix until uniform.

Wood Adhesive

(Polyvinyl Acetate)

Elvacet 89-100	95
Benzoflex 2-45	3
Water	2

Woodworking Adhesive

(Ethylene Vinyl Acetate/Resin)

EVA	30–60
Rosin Polyol Ester	30–60
Calcium Carbonate	0–50
Microcrystalline Wax	0–20
Plasticizer	0–5
Other Polymers	0–20
BHT	0.1–1

Edge Binding Adhesive

(Ethylene Vinyl Acetate/Resin)

EVA	40 (30–60)
Tackifying Resin	30 (30–50)
Whiting	30 (0–50)
BHT	0.5 (0.2–1)

Hot-Melt Furniture Edge Veneer Adhesive

(Ethylene Vinyl Acetate/Resin)

EVA-605	50.0
Arotap 546	30.0
Aroclor 5460	10.0
Multiwax 180-M	10.0
BHT	0.5

Plywood Core Adhesive

(Ethylene Vinyl Acetate)

EVA	40 (35–50)
Hydrogenated Rosin	50 (40–60)
Microcrystalline Wax	10 (0–20)
BHT	0.2 (0.1–1)

Hot-Melt Adhesive for Bonding Plywood

(Ethylene Vinyl Acetate/Resin)

Elvax 650	54
Super Nirez™ 5120	40
A-C® 680	5
Irganox® 1010	1

Properties:

Visc. @ 177 C	180,000 cps
Softening Pt. (Ring & Ball)	125 C

Water-Resistant Plywood Adhesive

(Water-Based)

Soya Protein	14.5
Pearl Starch	3.0
Stixso NN	16.0
Water	66.5

Procedure:

The simple mechanical dispersion of adhesives is prepared in the following manner: Add the water in a clean mixing tank. It is essential that the mixing vessel should not contain any residual silicate as this would tend to form agglomerates with the addition of the protein, and such lumps should not subsequently be dispersed. Use a high-speed mixer and sift the protein in slowly so as to prevent the formation of lumps. Add starch in a like manner. Continue mixing for approximately 5 min or until smooth. Add the **Stixso NN** sodium silicate rapidly with continuing agitation. Continue mixing for at least 15 min. During this period the viscosity of the adhesive will stabilize at its minimum of 30–50 Marsh funnel s.

When the adhesive is being prepared, aeration of the mix should be avoided to prevent foaming. Thinners are used if necessary to correct viscosity. Protein silicate adhesives set at lower temperatures than those normally used in adhesive processes. Excessive heat, therefore, is not necessary, so that only a small amount of heat is needed to control transfer.
Note:

The adhesives are made at the time of consumption, in batches that will be consumed in 4–8 h. If the temperature of the adhesive is maintained at a maximum of 70 F, the viscosity will remain within the desired range of workability for 24 h or longer. Normally, however, the temperature tends to increase in the glue pan on the machine where it is being applied.

Wood Patching Adhesive

	Formula No. 1	No. 2
	(Polyester/Resin)	
Base Component:	53.00	54.00
Dion DPM-1002	100.00	100.00
A-189	0.50	0.50
Dioctyl Phthalate	35.00	—
Chlorowax 500 C	—	25.00
Emersol® 132	1.80	1.80
Thixcin E	15.00	10.00
Super-Multiflex	110.00	110.00
Titanox® RA-NC	10.00	10.00
Thermax MT	1.00	1.00
Activator:		
Lead Peroxide (med. cure)	7.50	—
Lead Peroxide (slow cure)	—	7.50
Dioctyl Phthalate	6.75	6.75
Emersol® 232	0.75	0.75

Plastic Wood

Thixcin R has been used for years in conventional nitrocellulose dough to impart thixotropic body and nonslumping properties to wood-patching compounds. These compounds have good knifability and spreading properties free from curl.

The use of 6% of **Thixcin R** in acrylic resin or nitrocellulose gives excellent suspension without separation.

Wood Sealant

	Formula No. 1	No. 2
	(Black)	*(Tan)*
Resin		
Thiokol LP-32	100.0	100.0
Methylon 75109	—	5.0
Durez 10694	5.0	—
Multiflex MM	—	25.0
Icecap K	—	30.0
Titanox® RA-50	—	10.0
SRF #3	30.0	—
Attagel® 20	3.0	—
Cabosil® M-5	—	2.0
Sulfur	—	0.1
Stearic Acid	1.0	0.6
Aroclor 1254	—	35.0
Accelerator		
PbO$_2$ (med. grade)	7.5	7.5
PbO$_2$ (VFC, 0.5 μ)	—	7
Toluene	5.0	25.0

Panel/Subflooring Adhesive

(Acrylic Resin)

Rhoplex® LC-45	100.00
Triton® X-405	0.60
Composition T	1.22
Methocel® E-4M	0.20
Ethylene Glycol	4.00
Paraplex® WP-1	3.00
Tamol® 850	0.23
Camel Carb®	130.00
Nopco® NXZ	0.20
Water	4.00

Properties:

Pigment/Binder Ratio	2/1
Solids	82.2%

Note:
A suitable biocide is recommended for in-can preservation.

Panel Adhesive

(Solvent-Based, SBR/Resin)

Ameripol 4503	100.0
Cyanox 2246	1.0
Atomite®	400.0
Betaprene® BR-100	100.0
Super Sta-Tac® 80	100.0
Hexane	224.0

Procedure:

Ameripol 4503 requires high-speed and shear mixing in order to be dispersed in the organic solvent. Place the rubber into the solvent for a few hours. This causes a swelling effect in the rubber, which makes dispersion easier. After Step 1, the rubber and solvent are mixed with high speed and shear. During this time the resin and antioxidant can be added. Mixing is then continued until all of the resin has been dissolved and the adhesive is of a smooth consistency.

Properties:

Solids	75.7%
Visc.	Thick mastic

Vinyl Film-to-Particle Board Adhesive

Formula No. 1

(Rubber)

Hycar® 2600X138	50.0
Hycar® 2600X146	50.0
Nopco® 2271	35.0
Saniticizer 141	2.5
Acrysol® G-110	0.5
Propylene Glycol	25.0

No. 2

(Rubber)

Hycar® 2600X138	50.0
Hycar® 2600X146	50.0
Propylene Glycol	10.0
Acrysol® G-110	0.5

Coat the adhesives on the back, or printed, side of the vinyl sheet with a No. 36 wire-wound rod. Dry in a 49 C (120 F) oven or under an infrared heating lamp until the coating becomes clear. Press the coated vinyl onto the particle board at 49 C (120 F) at 0.17 MPa (25 psi).

Hot-Melt Vinyl Film-to-Wood Laminating Adhesive

(Ethylene Vinyl Acetate/Resin)

EVA-607	40.0
Arotap 546	25.0
Staybelite® Ester 10	15.0
Multiwax 180-M	20.0
BHT	0.5

Vinyl Film-to-Wood Adhesive

	Formula No. 1	No. 2	No. 3
		(Dry, Latex)	
Geon® 450X20	100.0	—	—
Geon® 460X2	—	50.00	50.00
Geon® 460X1	—	50.00	50.00
Methocel® MC (4000 std.)	2.0	—	—
Tetrasodium Pyrophosphate	—	0.10	—
Isophorone	—	—	2.00
Carbopol® 934	—	1.26	0.50
Trichloroethylene	—	5.00	—

Note:

Coat the fiberboard or wood substrate with these compounds using a #36 wire-wound rod. Laminate the vinyl film to the adhesive-coated substrate under pressure and slight heat. Then flash with an infrared lamp or place in a hot-air oven for 10–15 s. Then allow to dry and cure at room temperature.

Vinyl Film-to-Wood Adhesive with High Wet Tack

(Latex)

Ucar® 160	90.91
Cellosize® QP-15,000H	0.36
Gantrez AN 139 (20% sol'n.)	0.68
Xylene	5.00
Water	3.05
Preservative	0.10

Conductive Floor-Covering Adhesive

(Acrylic Resin)

Acronal 81 D	100.0
Caustic Soda (10%)	6.0
Nopco® NDW	0.3
Latekoll® D (12% sol'n.)*	8.4
Quartz Flour S 100	15.0
Derussol® VU 25 L	80.0
Balsam Rosin WW (70% sol'n. in toluene)	25.7

Uses:
Computer rooms and operating theaters.

---

*   The **Latekoll® D Solution** is produced as follows:

Water	47
Ammonia (25%)	5
Latekoll® D	48

Latekoll® D is incorporated into the ammoniacal water with vigorous stirring. The solution must be clear.

Flooring Adhesive

(Polyacrylate)

Texigel 13-037	502.0
Nopco® NXZ	5.0
Dibutyl Phthalate	45.0
Mix well. Premix:	
Texigel 13-302	12.5
Water	25.0
Add slowly with good mixing.	
Tektamer 38	0.3
Ammonia	7.5
Mix to smooth paste. Add:	
#1 Tripowhite Silica	470.0
Ruff Buff Silica	220.0

Properties:	
Total Solids	75%
Filler/Binder Ratio	2.5/1
W.P.G.	12.87 (12 lb 14 oz)

Pressure-Sensitive Flooring Adhesive

Formula No. 1

(Acrylic Resin)

Acronal 80 D	80
Acronal 7 D	20
Collacral VL	2
Aerosil® 200	2

No. 2

(Acrylic Resin)

Acronal 85 D	100
Collacral VL	2
Aerosil® 200	1

Note:

Collacral VL and Aerosil® 200 have a high thickening effect on the adhesive. This is necessary for imparting the required slump resistance to the adhesive beads.

No. 3

(Acrylic Resin)

Acronal 81 D	50
Acronal 880 D	50

No. 4

(Acrylic Resin)

Acronal 81 D	80
Hercolyn® 1151 (60% sol'n. in toluene)	20

Note:

Toluene causes the polymer particles to swell and thus thickens the adhesive. For this reason, the addition of thickening agent can be omitted or reduced.

The coat weight after drying should lie at 100–200 g/m^2.

Floor Tile Adhesives

	Formula No. 1	No. 2
	(White)	*(Natural)*
	(Oil-Based-Resin)	
W.W. Rosin (75% N.V.)	182	—
Pembro (75% N.V.)	—	270
Blown Soybean Oil Z-4	182	120
Rutile TiO_2	25	—
Snobrite Clay	550	600
Lead Naphthenate (24%)	10	10
Cobalt Naphthenate (6%)	2	2
Denatured Alcohol	175	147
Water	33	33

Procedure:

These adhesives are best prepared by thoroughly combining the resin cut (precut with a minimum amount of alcohol) and plasticizer in a heavy-duty mixer. The **Snobrite Clay** (and TiO_2 where used) can be added as rapidly as is consistent with good mixing efficiency. When the clay is well wetted out and dispersed, the naphthenate driers should be added, followed by the alcohol addition. The alcohol should be introduced slowly at first in order to avoid the formation of lumps, and then at a more rapid rate until all is added. The water should then be added at once and as soon as the mixture is uniform the compound should be packaged out to prevent as much loss of the alcohol as possible.

No. 3

(Latex/Resin)

Naugatex 2105	172.0
Nopco® 2271	262.0
Igepal® CO 880 (50% sol'n.)	13.5
12% Ammonium Caseinate*	71.0
Dioctyl Phthalate	11.0
Dowicide® G (20% sol'n.)	2.5
Age Rite Spar	2.0
6% CMC (Type 70, high visc.)	110.0
Atomite®	500.0

Procedure:

To obtain satisfactory dispersion of the ingredients it will be necessary to use an efficient blender, one which provides high internal shear, yet does not cause premature coagulation of the latex. Successful mixes on both a Baker-Perkins and a pony mixer have been made in the laboratory. The Baker-Perkins disperses the **Atomite®** more readily than the pony mixer, and in less time.

---

* *Preparation of 12% Casein Solution:* Casein (100), 28% ammonium hydroxide (25), water (694), 20% **Dowicide G** (15). Soak casein in one-half the cold water for 30 min. Commence stirring. Add ammonium hydroxide and heat to 60 C. Continue stirring until casein has dissolved (approximately 30 min); stop heating and add remainder of cold water.

For optimum stabilization of the latex system, the first eight ingredients should be stirred together thoroughly before addition of the **Atomite®**. A propeller-blade stirrer or pony mixer is adequate for the job. Once these eight ingredients are smoothly dispersed, the batch is then transferred to the Baker-Perkins; and while the kneading action is proceeding, small increments of **Atomite®** are added, allowing each portion to "wet-out" before a succeeding portion is added.

The kneading is continued until a smooth, homogeneous, agglomerate-free dispersion is obtained. This may be observed by making periodic sample smears of the cement during the mixing cycle. No tiny particles of **Atomite®** should be visible to the naked eye when dispersion is complete. A one-gallon mix takes approximately 40 min to disperse.

NOTE: Caution should be exercised to prevent excessive mixing, which will cause the cement to revert from a smooth to a coarse, "elastic" mix. This usually will happen if mixing is continued too long beyond the point where dispersion is complete.

The cement can be stored in glass jars or specially lined metal cans. Iron, copper, manganese, and other multivalent metallic ions have a deleterious effect on stability of the latex.

This adhesive has versatile working characteristics. Of very high viscosity, it can be removed from its container with a trowel or poured on the floor. However, once combed on a vertical surface it will not flow. It is "short" and "tender," and does not exhibit premature tack. Although worked freely back and forth with a notched trowel, it is not easily coagulated. The adhesive contains approximately 1% of a high flash-point solvent; therefore, fire hazard is at an absolute minimum. However, to insure maximum safety, working areas should be well-ventilated and the usual fire-prevention rules should be observed.

It is suggested that tiles be laid very soon after the adhesive is applied to the floor, and at least before the surface dries (approximately 15 min at 75 F). Initial "grab" is not so fast that tiles cannot be readjusted in position once they have been laid. Within one-half hour the tile is firmly held in position, and very good adhesion is obtained overnight. Excellent "set" is obtained in from three days to one week, depending on room temperature and humidity.

When under-floor repairs are necessary, tiles laid with this adhesive can be pried loose carefully, without damage to the tile. The adhesive, when fresh, can be removed from the tile with a damp cloth, or when dry, with common household solvents such as naphtha, gasoline, and benzine.

Wet Mastic Floor Tile Adhesive

Formula No. 1

(Acrylic Resin)

Rhoplex® N-580	98.20
Water	1.80
Propylene Glycol	3.00
Nopco® NXZ	0.09
Triton® X-405	2.50
Staybelite® Ester #10/Xylene (70/30)	76.70
Ammonium Hydroxide (10%)	2.00
Gold Bond® R Silica	80.55
Acrysol® ASE-60/Water (1/1)	5.00

Properties:

Solids	71.7%
Visc. (Brookfield HAT #4/0.3)	2,000,000 cps
Rosin/Binder Ratio	1/1
Pigment/Binder Ratio	1.5/1

Note:

A suitable biocide is recommended for in-can preservation.

	No. 2	No. 3	No. 4
	——— *(Acrylic Resin)* ———		
	Rhoplex® N-580	**Rhoplex® N-1031**	**Rhoplex® LC-67**
Emulsion (as supplied)	100.00	100.00	100.00
Triton® X-405	1.90	2.50	3.00
Nopco® NXZ	0.09	0.09	0.09
Acrysol® ASE-60/water (1/1)	1.90	5.00	5.00
Staybelite® Ester 10/Xylene			
(50% solids)	34.00	—	—
(70% solids)	—	76.70	92.86
Gold Bond® R Silica			
(300 mesh)	43.4	80.55	65.00
Ammonium Hydroxide (28%)	—	—	1.00
Sodium Hydroxide (10%)	6.00	2 00	—

Properties:

Pigment/Binder Ratio	0.79/1	1.5/1	1/1
Resin/Binder Ratio	0.31/1	1/1	1/1
pH	8.0	8.0	8.0
Solids, %	62.5	71.4	74.2
Visc., cps	50,000	2,000,000	2,000,000

Emulsion-Type Ceramic Tile Mastic

Formula No. 1

(Natural Latex)

Triethanolamine	30
Oleic Acid	56
TPO #2	174
TPO #3	58
Atomite®	250
Snobrite Clay	200
Natrosol® 250 H (3% in water)	253
Naugatex 2105 (62% N.V.)	80

	No. 2	No. 3
	(Natural/Resin)	*(White/Resin)*
Triethanolamine	30	30
Oleic Acid	56	56
Piccovar® AP-25	117	117
Piccolyte® S-25	109	109
Rutile TiO_2	—	50
Atomite®	250	250
Snobrite Clay	200	200
Natrosol® 250 HR (3%)	263	250
Pliolite 5356 (69% N.V.)	71	71

Procedure:

The triethanolamine and the oleic acid should be added first into a heavy-duty mixer (e.g., double sigma-blade Baker-Perkins) and mixed until completely reacted into a clear gel-like soap.

The liquid and the solid resins are then added (heating if necessary) with continuous mixing. This is followed by the pigments. The above combination should be mixed until uniform.

The major portion of the hydroxyethyl cellulose solution is then added and the agitation continued until a homogeneous oil-in-water emulsion is formed.

The remaining portion of the hydroxyethyl cellulose solution is combined with the latex binder in order to give the latex the maximum protection while it is added to the pigmented resin solution. This latex–hydroxyethyl cellulose solution combination should be carefully added as the final step.

The agitation should be stopped as soon as the latex is thoroughly incorporated. The mastic should be packaged in cans that have been lined to prevent rust and latex coagulation.

No. 3

(Acrylic Resin)

A	Acronal S 400	100.0
	Plastilit 3060	5.0
	Sodium Gluconate	0.5
	Nopco® NXZ	1.0
	Preservative CA 24	0.3
	Walocel® MW 40,000 PV (2% aq. sol'n.)	17.5
	Ammonia (25%)	0.3
	Rohagit® SD 15	1.5

B White cement

100 parts Component A is mixed with 150 parts Component B on the building site. The mix must be stirred until it is homogeneous. Tests must be carried out in every case to ensure that the cement is compatible with Component A.

Procedure:

The **Acronal 290 D** or **295 D** is placed in the vessel, and the **Plastilit 3060** and other additives incorporated in the sequence given. The ingredients are added in portions, and the mixture stirred after each addition until smooth.

All mixers with a suitable stirrer system can be used to produce these high-viscosity adhesives. Planetary and turbulent-flow mxiers are particularly suitable.

The addition of a preservative is recommended to protect the adhesives from attack by microorganisms. Testing and monitoring must be carried out to check for suitability.

Note:

This is a 2-part, flexible, water-resistant adhesive.

Ceramic Tile Adhesive

Formula No. 1

(Acrylic Resin)

Acronal 290 D (295 D)	100.0
Mineral Spirit (145–200 C)	5.0
Nopco® NXZ	0.5
Preservative CA 24	1.6
Pigment Disperser A	0.3
Sodium Polyphosphate	0.2
Latekoll® D (4% sol'n.)*	120.0
Juraperle® MH	600.0
Methanol	10.0

No. 2

(Acrylic Resin)

Acronal 290 D (295 D)	100.0
Plastilit 3060	10.0
Nopco® NXZ	0.5
Preservative CA 24	0.9
Pigment Disperser A	0.1

Sodium Polyphosphate	0.1
Latekoll® D (4% sol'n.)*	37.5
Juraperle® MH	300.0
Methanol	5.0

Note:

Mineral spirit reduces the film-forming temperature from 23 C to 0–5 C. **Nopco® NXZ** is a defoamer for aqueous polymer dispersions. **Preservative CA 24** protects against attack by microorganisms. **Pigment Disperser A** and sodium polyphosphate are surfactants to improve the homogenity and storage stability of highly filled adhesives. **Latekoll® D Solution,** a protective colloid, is used to control the consistency, open time, and spreading properties of the adhesive. **Juraperle® MH** is a finely crystalline calcite with a particle size distribution of 0–100 μm and a specific surface of 0.23 m^2/g. It reinforces the adhesive film. Methanol imparts some degree of frost resistance, i.e., adhesive that has frozen can be rendered reusable after thawing out and careful stirring.

---

* The **Latekoll® D Solution** is produced as follows:

Water	82
Ammonia (25%)	2
Latekoll® D	16

The water and the ammonia are placed in the vessel, and **Latekoll® D** added at a high stirrer speed. The solution must be clear before further processing is undertaken.

	No. 3	No. 4
	(Oil-Based/Resin)	
Jellybean Oil	290	—
Blown Soybean Oil (z-6+)	—	310
Neville LX-1000 (70% N.V.)	120	128
Rutile Titanium Dioxide	50	50
Snobrite Clay	350	350
Snowflake White	250	300
Cobalt Drier (6%)	3	3
Denatured Alcohol	135	94
Water	7	4

No. 5
(Water-Based, Latex/Resin)

Betaprene® BR-100	94.50
Naphthenic Oil (1200 SUS visc. @ 37.7 C)	25.50
VM&P Naphtha	25.00
Aconew® 500	8.75
Ammonium Hydroxide/Water (50/50)	16.00
Tylac® 97-454 (51%)	192.00
Hydrite R	120.00

Properties:
Solids 70%

Procedure:

Dissolve **Betaprene® BR-100** in the naphtha and oil. When all the resin is dissolved, add **Aconew® 500** to the solution, mix thoroughly. Add the diluted ammonium hydroxide to resin solution, *slowly*, with constant mixing. *Slowly* introduce the **Tylac® 97-454** with constant shear. After the latex has all been introduced, the **Hydrite R** can be added with high-shear mixing.

No. 6
(Water-Based)

Experimental Emulsion E-1997	210.0
Propylene Glycol	10.0
Water	70.0
Tamol® 731	5.0
Urea	30.0
Defoamer	1.0
Duramite	500.0
Acramin Clear Concentrate NS2R	14.0

Properties:

Solids	76.0%
Pigment/Binder Ratio	5/1
Shelf Stability, 1 mo. 50 C	Pass
Freeze/Thaw Stability	5+ cycles
Approximate Open Time	1 h
ANSI #1	40+ lb

Note:

A suitable biocide is recommended for in-can preservation.

No. 7

(Solvent Based, SBR/Resin)

Ameripol SBR-4503	100.0
Cyanox 2246	1.5
Hydrite R	400.8
Atomite®	133.5
Betaprene® AC-140	70.0
Super Sta-Tac® 100	100.0
Paraffin Wax (m.p. 150 C)	22.9
Hexane	343.6

Procedure:
Swell SBR elastomer in hexane. Under high-speed agitation, add **Super Sta-Tac® 100** and **Betaprene® AC-140**; when dissolved add wax. Continue mixing until all the resins and wax are dissolved. Last, add fillers.

Properties:

Solids	70.7%
Open Time	≥ 2 h

	No. 8	No. 9
	(Latex/Resin)	
Hydrocarbon Resin	100.0	100.0
Mineral Spirits	67.0	67.0
Oleic Acid	3.0	3.0
Potassium Hydroxide (10%)	7.5	7.5
Casein (15% sol'n.)	20.0	20.0
Clay Slurry (65%)	233.0	388.0
NR Latex (HA or LA-TZ type, 60%)	167.0	167.0
Water	18.0	18.0
Zinc Diethyl Dithiocarbamate (50%)	2.0	2.0
Celacol HPM	0–4	0–4

No. 10

(Water-Based/Polyacrylate)

Water	175
Texigel 13-302	9
Nopco® NDW	4
Dibutyl Phthalate	18
Preservative	1
Mix; add:	
Natrosol® 250 HR	10
Disperse then add:	
Ammonia (28%)	5
Mix until smooth:	
Texigel 13-037	400
Picconol® A-200	40
Mix thoroughly:	
#10 white	275
Gammasperse 255	275

No. 11

(Water-Based/Polyacrylate)

Water	360
Texigel 13-302	23
Foamaster VF	4
Carsonon N-50	5
Dibutyl Phthalate	2
Preservative	1
Mix thoroughly:	
Natrosol® HR	8
Disperse:	
Ammonia (28%)	5
Mix until smooth:	
Texigel 13-037	110
Mix thoroughly:	
Omyacarb #10	1000

No. 12

(PVAc)

Portland Cement Type 1	50.00
Silica Sand (100–300 μ)	50.00
Vinnapas Redispersible Powder RE 526 Z	2.50
Tylose MHB 30,000 yp	0.25

No. 13

(Latex/Resin)

Natural Rubber Latex (sodium pentachlorphenate preserved—60%)	100
Polyvinyl Acetate Latex (50%)	120
Surfactant (10% sol'n.)	10
Gypsum	200–250
Bentonite Clay	20
Tackifying Resin Dispersion	20
Water	as required

No. 14

(SBR)

Rubber SBR-1009	10.0
Panarez 6-210	2.0
Heptane	35.0
Panarez 6-210	23.0
Thixcin R	2.5
Hydrite Clay	27.5

Procedure:

Mill **Rubber SBR-1009** and **Panarez 6-210** (first portion) on a rubber mill. Add heptane and let swell overnight. Then mix on a Cowles Dissolver. Heat to 140 F (60 C). Heat **Panarez 6-210** (second portion) to 210 F (99 C) to melt it and then stir into above mix. Cool to 140 F (60 C) and add **Thixcin R.** Cool overnight to room temperature. Add **Hydrite Clay** and mix in a change can mixer.

No. 15

(Styrene-Butadiene/Resin)

Solprene 303	100
Solprene 375	25
Stanwhite 325	100
Dixie Clay	200
Dymerex	50
Antioxidant 2246	2
Hexane	100

Vinyl Tile Adhesive

(Latex/Resin)

Terpene-Phenolic Resin	100.0
Pale Liquid Coumarone Resin	50.0
Process Oil	50.0
Oleic Acid	22.5
Mineral Spirits	25.0
Potassium Hydroxide (10% sol'n.)	40.0
Casein (15% sol'n.)	20.0
Water	75.0
Zinc Diethyl Dithiocarbamate (50% disp.)	4.0
NR Latex (HA or LA-TZ type, 60%)	167.0
Clay	100.0
Celacol HPM	as required

Floor and Wall Tile Mastic

(Styrene-Butadiene/Resin)

Ameripol 1013	25
Ameripol 1009	75
Calcium Carbonate	40
Paragon	120
Cumar® MH 2½	45
Picco® LTP-135	50
Antioxidant 2246	2
Heptane	200

Synthetic Wall-and-Floor-Covering Adhesive

(Acrylic Resin)

Acronal 81 D	100.0
Caustic Soda (10%)	6.0
Nopco® NDW	0.3
Latekoll® D (12% sol'n.)*	8.4
Fine Chalk	40.0
Quartz Flour S 100	30.0
Balsam Rosin WW (70% sol'n.)**	35.0

---

* The **Latekoll® D Solution** is produced as follows:

Water	47
Ammonia (25%)	5
Latekoll® D	48

Latekoll® D is incorporated into the ammoniacal water with vigorous stirring. The solution must be clear.

** **Balsam Rosin WW** is dissolved in a mixture consisting of 2 parts of ethyl acetate and 1 part of methyl acetate.

In order to achieve uniform distribution of the resin solution in the adhesive, it is necessary to use high-speed mixers for producing one-side adhesives. Turbine and planetary mixers are particularly suitable for this purpose.

Note:

It is recommended that the dispersion-based adhesives be blended with a preservative in order to prevent them from the attack of microorganisms.

Rigid, Expanded Plastic Tiles Adhesive

(Polyvinyl Propionate)

Propiofan 5 D or **70 D**	35
Latekoll® D (4% sol'n.)	15
Carborex® 20	65

Note:

This adhesive must be of paste consistency for ease of application in overhead work.

Floor and Wall Tile Mastic

Formula No. 1

(Rubber/Rosin)

Reclaimed Rubber	16.39
Rosin Still Residue	24.69
Dixie Clay	8.23
Calcene TM	8.23
Asbestos 7TO6	8.23
Neozone D	0.24
Touene	15.30
n-Hexane	18.70

Procedure:

Above is mixed in duplex mixer at 230 F. The reclaimed rubber is then added and masticated for 15 min, first at 33 and later at 84 rpm. Next, the rosin derivative is added to the mill, and after 15 min of mixing the mill is cooled to 190 F. The speed is reduced to 50 rpm and the clay is then added slowly. After another 15 min of mixing, the mill is cooled to 175 F. The mill speed is then reduced to 33 rpm and the calcium carbonate and antioxidant are added slowly to the batch.

After 5 more min of mixing, the toluene is added over an 8–10 min. period. The mill is cooled to 100 F, at which time the asbestos is added. Following 10 min of mixing, the *n*-hexane is added slowly to the mix. After solvent addition, the mastic is allowed to mix for an additional 5 min.

No. 2

(Water-Based, SBR/Latex)

Water	180
SBR 2002 Latex (48% T.S.)	352
Whiting	336
Titanium Dioxide	5
Carbopol® 934	5
Ammonium Hydroxide	to mucilage pH 9.5

Procedure:

Disperse the Carbopol® 934 in the water with moderate mixing and neutralize to pH 9.5 with ammonium hydroxide. Slowly add the latex to this preneutralized Carbopol® 934 mucilage with moderate mixing, such as with a Lightnin' propeller-type mixer. Add the pigment, very slowly, with moderate agitation and stir until the product is smooth.

Note:

An important step for the successful preparation of this product is that the **Carbopol® 934** solution must be neutralized to a pH of 9–9.5 before the latex is added. A lower pH will cause coagulation of the rubber and/or flocculation of the pigments.

No. 3

(Styrene-Butadiene/Resin)

Solprene 411P	100
Picco® 6140	75
Picco® 6100	75
Clay	100
Whiting	100
Antioxidant	3

No. 4

(Styrene-Butadiene/Resin)

Solprene 411P	75
Cross-Linked Polymer	25
Picco® 6100	25
Escorez® 2101	50
Clay	200
Whiting	400
Naphthenic Oil	20
Antioxidant	2

	No. 5	No. 6
	(Acoustical Tile, Resin)	(Cove-Base Cement, Resin)
Tolrez	247	181
Pembro (75% N.V.)	224	390
Kaolloid Clay	600	550
VM&P Naphtha	19	—
Denatured Alcohol	67	27
Water	17	13

Procedure:

Thoroughly combine the tall-oil pitch and resin cut in a heavy-duty mixer. The **Kaolloid Clay** can then be added as rapidly as is consistent with good mixing efficiency. When the clay is well wetted out and dispersed, the alcohol should be introduced, slowly at first to avoid the formation of lumps, then at a more rapid rate until all is added. The water should then be put in at once and as soon as the mixture is uniform it should be packaged out to prevent as much loss of the alcohol as is possible. The purpose of the alcohol or alcohol and water addition is to produce the buttery characteristics so necessary for good application.

Note:

The acoustical tile adhesive should be stiff enough in consistency to adequately support the weight of the tiles and have the proper balance between adhesion and cohesion in order to have a permanent bond to the tile and the substrate.

The cove-base cement requires a greater ratio of resin cut to the tall-oil pitch (plasticizer) in order to give the extra adhesive strength and rapid set needed to effectively hold this "L-shaped" tile in position indefinitely even under the pressures of dimensional changes of the tile caused by temperature variations.

Emulsion-Type Wall Tile Mastic

(Latex/Resin)

	No. 1 (White)	No. 2 (Natural)
Triethanolamine	26	26
Oleic Acid	49	49
Piccolyte® S-25	199	—
Piccovar® AP-33	—	201
Atomite®	800	800
Natrosol® 250-H (3% in water)	217	217
Naugatex 2105 (62% N.V.)	36	36

Procedure:

The triethanolamine and the oleic acid should be added first into a heavy-duty mixer and mixed until completely reacted into a clear gel-like soap.

The preheated plasticizer–tackifier is then added (heating if necessary) with continuous mixing. This is followed by the pigment and mixed until uniform.

The major portion of the hydroxyethyl cellulose solution is then added and the agitation continued until a homogeneous oil-in-water emulsion is formed.

The remaining portion of the hydroxyethyl cellulose solution is combined with the latex binder in order to give the latex the maximum protection while it is added to the pigmented solution. This latex–hydroxyethyl cellulose solution combination should be carefully added as the final step.

The agitation should be stopped as soon as the latex is thoroughly incorporated. The mastic should be packaged in cans that have been lined to prevent rust and latex coagulation.

Dispersion Adhesives for Bonding PVC Floor Coverings

Formula No. 1

(Acrylic Resin)

Acronal 80 D	100.0
Caustic Soda (10%)	6.0
Nopco® NDW	0.3
Latekoll® D (12% sol'n.)*	11.7
Fine Chalk	15.0
Quartz Flour S 100	35.0
Balsam Rosin WW (70% sol'n. in toluene)	21.0

No. 2

(Acrylic Resin)

Acronal 81 D	100.0
Caustic Soda (10%)	6.0
Nopco® NDW	0.3
Latekoll® D (12% sol'n.)*	8.4
Fine Chalk	18.0
Quartz Flour S 100	42.0
Balsam Rosin WW (70% sol'n. in toluene)	25.0

---

* The **Latekoll® D Solution** is produced as follows:

Water	47
Ammonia (25%)	5
Latekoll® D	48

Latekoll® D is incorporated into the ammoniacal water with vigorous stirring. The solution must be clear.

Dispersion Adhesives for Bonding PVC Foam-Backed Floor Coverings

(Acrylic Resin)

Acronal 81 D	70.0
Caustic Soda (10%)	4.2
Acronal 355	30.0
Nopco® NDW	0.3
Latekoll® D (12% sol'n.)*	11.7
Fine Chalk	18.0
Quartz Flour S 100	42.0
Balsam Rosin WW (70% sol'n. in toluene)	10.0

---

* The **Latekoll® D Solution** is produced as follows:

Water	47
Ammonia (25%)	5
Latekoll® D	48

Latekoll® D is incorporated into the ammoniacal water with vigorous stirring. The solution must be clear.

Nonlaminated Wallpaper Underlay Adhesive

(Water-Based, Polyacrylate)

Propiofan 5 D	**20**
Chalk (finely divided)	20
Latekoll® D (4% sol'n.)*	60

---

* The **Latekoll® D Solution** is produced as follows:

Water	82
Ammonia (25%)	2
Latekoll® D	16

The **Latekoll® D** is incorporated into the ammoniacal water under high-speed stirring. The resultant solution must be clear.

Note:

A lambswool roller can be used to apply the adhesive.

Wall Paper Coatings

Formula No. 1

(Ground Coat, Water-Based)

Clay	100.0
Water	80.0
Calgon S	0.3
Texicryl 13-104	17.0

No. 2

(Printing Coat, Water-Based)

Clay	100.0
Calgon S	0.3
Water	79.0
Casein (15% sol'n.)	30.0
Texicryl 13-104	16.0
Dye	as required

Vinyl Wallcovering Adhesive

Formula No. 1

(Water-Based, PVAc)

Water	33.0
Tamol® 731	1.0
Tammsco Velveteen R	35.0
Douglas Starch Thin XF	5.0
Cellosize® QP 30,000	0.5
Polyco 2113	25.0
SA-13	0.5

Procedure:

Add first five ingredients in order shown. Cook at 190–195 F to paste while agitating.   Cool to 150 F.   Add **Polyco 2113** and **SA-13**. Continue to agitate while cooling. Stir until smooth.

Properties:

Solids	55%
Visc.	Paste

No. 2
(Water-Based, Latex)

Ucar® 160	40.55
ASP®-400	14.60
Diethylene Glycol	4.06
Butyl Carbitol®	2.23
Cellosize® QP-52,000H	1.22
Water	34.91
Borax (10% aq. sol'n.)	2.43

Procedure:

Premix all ingredients except borax solution and latex. Stir premix into the latex, followed by the borax solution.

Pressure-Sensitive Adhesives for Insulating Materials
Formula No. 1
(Acrylic Resin)

Acronal 80 D or 81 D	70
Acronal 880 D	30

No. 2
(Acrylic Resin)

Acronal 85 D and, if necessary,	100
Collacral VL	1

Note:

For the casting process, the adhesive must have a comparatively low viscosity. For this reason, **Acronal 85 D** is a suitable base product for this application.

No. 3
(Acrylic Resin)

Acronal 880 D	50
Acronal 81 D	50

Note:

Acronal 880 D or **V 205** are suitable base products for polyurethane foams.

No. 4

(Acrylic Resin)

Acronal V 205	80
Acronal 85 D	20

Note:

The coat weight after drying should lie at 80–100 g/m^2.

Wall Primer Sealer

(Water-Based)

Pigment grind:

Water	266.20
Phenyl Mercuric Acetate (100%)	0.25
Cellosize® QP-15000	4.00
Tamol® 731	4.60
Potassium Tripolyphosphate	1.00
Tergitol® NP-10	2.20
Colloid 581 B	0.90
Ethylene Glycol	9.30
Titanium Dioxide	75.00
Aluminum Silicate	150.00
Calcium Carbonate	150.00

Let down:

Ucar® 505	428.80
Texanol	17.80
Colloid 581 B	1.90
Ammonium Hydroxide (28%)	1.80

Fibrated Plastic Roof Cement

	Formula No. 1	No. 2
Sohio Grade B Cutback	24.00	24.00
RD4540	0.83	0.83
Mix 2 min.		
Attagel® 36	6.00	6.00
Mix until completely gelled (5–10 min).		
Sohio Grade B Cutback	44.92	44.92
Mix until homogeneous (5 min).		
Pulpex®-P AD-H	1.25	—
Polyolefin Fibers	—	1.25
Emtal® 42	23.00	23.00
Mix until homogeneous (10 min).		

Properties:

Softening Pt. (Ring & Ball)	167 F
Needle Penetration	35 @ 77 F
Cutback Solids	62%/wt.
Saybolt Furol Visc.	120 s @ 122 F

	No. 3	No. 4
Sohio Grade B Cutback	24.0	24.0
Arquad® 2HT-75	0.8	0.8
Mix in double planetary mixer for 2 min (7 rpm).		
Attagel® 36	6.0	6.0
Mix until completely gelled (5–10 min).		
Sohio Grade B Cutback	44.1	44.1
Mix until homogeneous (5 min).		
Kay-O-Cel #10	2.0	2.0
Emtal® 42	23.0	—
Calcium Carbonate	—	23.0
Concentrated Phosphoric Acid	0.1	0.0
Mix until homogeneous (10 min).		

Properties:

Softening Pt. (Ring & Ball)	167 F
Needle Penetration	35 @ 77 F
Cutback Solids	62%/wt.
Saybolt Furol Visc.	120 s @ 122 F

Mastic Asphalt-Roof Cement*

Formula No. 1

Asphalt Cutback	40.0
Arquad® 2C-75	1.3
Mix 2 min at slow speed.	
Attagel® 36	8.0
Mix 10 min at medium speed.	
Asphalt Cutback	30.0
Wollastonite	20.7
Mix 10 min at medium speed.	

Properties:

Clay/Surfactant Ratio	6/1
Mineral Content	28.7%/wt.
Softening Pt. (Ring & Ball)	150–170 F
Needle Penetration	25–40
Cutback Solids	60%/wt.
Saybolt Visc.	140–190 S @ 120 F

---

*   Formulated to meet: Federal Specifications SS-C-153C
ASTM Specification D-2822-75

	No. 2	No. 3
Sohio Grade B Cutback	24.0	24.0
Arquad® 2HT-75	0.8	0.8
Mix in double planetary mixer for 2 min (7 rpm).		
Attagel® 36	6.0	6.0
Mix until completely gelled (5–10 min).		
Sohio Grade B Cutback	44.1	44.1
Mix until homogeneous (5 min).		
Kay-O-Cel #10	2.0	2.0
Emtal® 42	23.0	—
Chem Carb® 77	—	23.0
Concentrated Phosphoric Acid	0.1	0.0
Mix until homogeneous (10 min).		

Properties:

Softening Pt. (Ring & Ball)	167 F
Needle Penetration	35 @ 77 F
Cutback Solids	62%/wt.
Saybolt Furol Visc.	120 s @ 122 F

Roof Coating

Natrosol® 250MR (3%)	147.8
Water	16.8
Tamol® 850	5.0
KTPP	1.5
Ethylene Glycol	25.6
Nopco® NXZ	2.0
Ti-Pure R-960	73.9
Duramite	443.4
Kadox 515	49.3

Grind on high-speed disperser for 15 min. Then let down with the following ingredients:

Rhoplex® EC-1685	431.4
Texanol	7.3
Skane M-8	2.2
Nopco® NXZ	2.0
Ammonium Hydroxide (28% ammonia)	1.0
Water	17.9

Grind for an additional 10 min.

Properties:

Pigment to Binder Ratio	2.08/1
lb/gal	12.28
PVC	43.0%
Nonvolatile by weight	68.9%
Nonvolatile by volume	58.2%

Note:

This formula should be applied in two coats for a total dry-.film thickness of 30 mils. This should yield a coverage of approximately 30 ft^2/gal.

Rubberized Asphalt Joint Sealer

(Styrene-Butadiene)

Solprene 1205C	100.0
Airblown Asphalt	1320.0
Philrich 5	330.0
Zinc Oxide	3.3
Stearic Acid	1.0
Sulfur	5.3
BTDS	1.3
Bismate	1.3

Procedure:

Dissolve **Solprene 1205C** in molten asphalt and oil at 350 F. Add curatives and mix for 1 h at 350 F. The product is pourable when hot but upon cooling it becomes elastomeric.

Asphalt Concrete Joint Sealant

Kraton® 1101	100.0
Asphalt (vacuum residue)	180.0
Shellflex® 881	60.0
Picco® LTP 100	60.0
Antioxidant	1.0

Note:

Designed according to Fd. Spec. SS-S-00164.

Troweling Weather-Barrier Mastic

(Acrylic/Latex)

Ucar® 153 (55.5% T.S.)	685.92
Super Imperse Blue B	1.04
Calgon	3.14
Calidria® SG-210	12.41
Mineralite 3X	143.31
TiPure R-901	9.90
Kingsley Clay	47.73
Chlorowax 70	22.32

Antimony Oxide	1.88
Dricalite SA-3	69.98
Flexol® TCP	62.99
PMA-30	2.24
Colloid 581B	2.69

Procedure:

Add in order shown. Formulation is mixed 2 h in a 1-gal Baker-Perkins sigma blade-type mixer, at 79 rpm. Enclosed mixer minimizes moisture loss.

Consistency letdown, for brush or spraygun grade, can be made with water. It is suggested that the extra water be added toward the end of the mixing cycle.

Joint Cement Compounds

Formula No. 1

(Polyvinyl Acetate)

No. 1 White	500
AA Mica	430
1K Mica	25
Gelva 702	27
Gelvatol 20-30 BP	9
Sodium Acetate	2
Carboxymethyl Cellulose 7HSX	6
Dowicide® A	1

No. 2

(Casein)

No. 1 White	500
AA Mica	410
1K Mica	25
Daxad 11	1
Ben-a-gel EW	3
Sodium Bicarbonate	9
Sodium Carbonate	6
Dowicide® A	1
Casein (80 mesh)	45

No. 3

(Water-Based, Polyvinyl Acetate)

Triethanolamine	2.0
Oleic Acid	4.0
Water	226.0
Methocel® 65 HG (4000 cps 3% in water)	33.0
Polyvinyl Acetate (55% N.V.)	136.0
Duramite	900.0
Snobrite Clay	150.0
1K Mica	50.0
Troysan PMA-30	0.3
Mineral Spirits	26.0

Procedure:

Mix in suitable equipment the triethanolamine and oleic acid until they are completely reacted. This is indicated by the formation of a clear gel. Water, **Methocel®** solution, and polyvinyl acetate emulsion are then added to the soap and thoroughly mixed. The pigments should be added slowly, followed by the preservative and the mineral spirits. The compound should be quite smooth and homogeneous throughout at the end of the mixing period. The material should be packaged in cans that have been lined to prevent rust and latex coagulation.

Spackling Compounds

Formula No. 1

(Polyvinyl Acetate)

No. 1 White	962
Gelva 702	22
Gelvatol 20-30 BP	7
Sodium Acetate	2
Carboxymethyl Cellulose 7HSX	6
Dowicide® A	1

No. 2

(Casein)

No. 1 White	950
Daxad 11	1
Ben-a-gel EW	4
Sodium Bicarbonate	9
Sodium Carbonate	6
Dowicide® A	1
Casein (80 mesh)	29

No. 3

(Water-Based, Polyvinyl Acetate)

Triethanolamine	2.0
Oleic Acid	4.0
Water	135.0
Methocel® 65 HG (4000 cps 3% in water)	125.0
Polyvinyl Acetate (55% N.V.)	136.0
Duramite	1100.0
Troysan PMA-30	0.3
Mineral Spirits	26.0

Procedure:

Mix in suitable equipment the triethanolamine and oleic acid until they are completely reacted. This is indicated by the formation of a clear gel. Water, **Methocel®** solution, and polyvinyl acetate emulsion are then added to the soap and thoroughly mixed. The pigments should be added slowly, followed by the preservative and the mineral spirits. The compound should be quite smooth and homogeneous throughout at the end of the mixing period. The material should be packaged in cans that have been lined to prevent rust and latex coagulation.

Concrete Adhesives

	Formula No. 1	No. 2	No. 3	No. 4
		(Epoxy/Resin)		
A **LP-3**	100	100	100	150
Mortar White Silica				
(HDS-100)	80	80	—	—
Hydrite 121	—	—	140	175
DMP-10	—	10	—	—
EH-330	20	10	20	24
Toluene	—	—	65	60
B **Epon® 828** or equivalent	200	200	200	200
Hydrite 121	—	—	105	185
Toluene	—	—	5	19
Properties:				
Mixing Ratio	100/100	100/100	1/1	1/1
	by wt.	by wt.	by vol.	by vol.
Brushing Life (qt)				
@ 80 F, h	0.2	0.3	1.4	1.4
Set* Time (10 mil film)				
@ 80 F, h	0.8	1.0	3.5	2.0
Cure** Time (10 mil film)				
@ 80 F, h	7	8–24	8–24	8–24
Wt./gal	8.4	10.1	8.3	10.8
Nonvol. % wt.	55	70	56	72
Nonvol., % vol.	46	57	40	57.5
Mix Ratio A/B/C pbw	1/2	1/1	100/286	100/222
Pot Life, (spray) h	4	6	3	6
Dry-to-handle, h	2.5	1.5	2	4–5
Dry-to-walk-on, h	2–3	2–3	15–24	24–48
Min. Cure Time, days	3	3	3	4
Optimum Cure, days	7	7	7	7–10

Uses:

Formula No. 1 is a clear general-purpose coating; No. 2 is a general-purpose maintenance coating; No. 3 is an aluminum maintenance coating; and No. 4 is a sprayable, coal tar-modified LP/EP coating.

---

* Time after fresh concrete will not bond
** Time for adhesive to cure sufficiently to allow normal use of structure

Concrete Sealer

Formula No. 1

Water	63.860
FC-120 (25% as supplied)	0.024
SWS-211	0.036
Butyl Carbitol®	4.800
Ethylene Glycol	2.000
Dibutyl Phthalate	2.400
Tributoxy Ethyl Phosphate	0.300
Formalin (37%)	0.150
Adjust to pH 7.5 with ammonia.	
Rhoplex® WL-91 (41.5%)	36.140

Properties:
Density @ 25 C   8.42 lb/gal

No. 2

Water	39.76
FC-120 (25% as supplied)	0.04
SWS-211	0.06
Butyl Carbitol®	8.00
Ethylene Glycol	2.50
Dibutyl Phthalate	4.00
Tributoxy Ethyl Phosphate	0.50
Formalin (37%)	0.15
Adjust to pH 7.5 with ammonia.	
Rhoplex® WL-91 (41.5%)	60.24

Properties:
Density @ 25 C   8.47 lb/gal

No. 3

(Acrylic Resin)

Shanco 100	4.5
Neo Cryl B-700	10.5
Mineral Spirits	85.0

Procedure:

Add the resin and polymer to the solvent under good agitation. If desired, the solvent may be heated to faciliate solution. Should a defoamer be desirable, 250 ppm silicone oil (350 cstk) may be added.

Architectural Sealing Tape

	Formula No. 1	No. 2	No. 3	No. 4
		(Polybutene)		
EX-214	100	100.0	100	—
Bucar 5000NS	—	—	—	100
Indopol® H-100	100	140.0	100	100
Camel Carb®	50	—	25	150
Asbestine 3X	200	50.0	125	—
HiSil 233	20	20.0	20	—
Ti-Pure® R-900	20	20.0	20	20
Calcene TM	—	500.0	250	—
NBC	—	0.5	—	—
Indopol® H-1900	—	—	25	—

Procedure:

Mixing procedure for Baker-Perkins, cold water on.

0 min	Charge to mixer and blend all ingredients *except* polymer, $^1/_3$ **Indopol®** and $^1/_2$ filler.
1 min	Add polymer.
10 min	Add $^1/_4$ filler and remaining **Indopol®**.
15 min	Slowly add remaining filler.
30 min	Dump.

No. 5

(Polybutene)

Butyl Elastomer	6.00
Amoco® H-300	12.00
Vinyl Toluene/Vegetable Drying Oil Copolymer	3.00
Thixotrope	0.39
Calcium Carbonate	39.17
Talc	25.01

Titanium Dioxide	4.02
Mineral Spirits	9.96
Cobalt Drier (6%)	0.03
Antioxidant	0.12
Phenolic Resin (70% in xylene)	0.30

Procedure:

In the laboratory the sealant is prepared in a sigma blade mixer equipped with compression ram and a jacket for steam and cold water using the following steps:

Start	Steam on to 120 C (250 F); add thixotrope.
2 min	Add polybutene, antioxidant, and phenolic resin.
5 min	Steam off; incrementally add fillers and pigment.
15 min	Cold water on; add rubber; lower ram.
30 min	Add copolymer incrementally.
35 min	Add mineral spirits and cobalt drier incrementally.
50 min	Mix.
55 min	Dump.

No. 6

(Rubber)

Hycar® 4054	100
Hydral 710	200
N-550 Black	40
TCP Plasticizer	50

No. 7

(Nondrying, Polybutene)

Amoco® H-300	24.01
Amorphous Polypropylene Homopolymer	5.05
Butyl Rubber	1.72
Clay	16.37
Calcium Carbonate	44.05
Diatomaceous Silica	4.00
Cotton Fiber	4.80

Procedure:

This is compounded in a sigma blade mixer. Amorphous polypropylene is premixed with twice its weight of polybutene to facilitate complete dispersion. Additions to the mixer are made in the order of listing in the formula. The entire mass is then mixed for 1 h after the last addition.

	No. 8	No. 9	No. 10
		(Polybutene)	
Amoco® H-300	18.90	19.90	32.0
Polyisobutylene Rubber	18.90	19.90	—
Butyl Rubber	5.25	—	8.0
Calcium Carbonate	35.05	35.05	35.0
Platelet Talc	6.50	7.50	—
Attapulgas Clay	11.00	12.25	16.0
Diatomaceous Silica	2.40	3.40	5.0
Titanium Dioxide	2.00	2.00	4.0

Procedure:

In the laboratory the rubbers and polybutene are mixed in a dual-arm sigma blade mixer for approximately 20 min. Dry ingredients are incrementally added in the order listed while mixing. The sealant is dumped, rested 24 h, and extruded.

	No. 11	No. 12	No. 13
	(Polybutene/Butyl Rubber)		
Polysar Butyl XL-20	100	100	100
HiSil 233	—	—	60
N-770 (SRF)	120	—	—
Hard Clay	50	40	40
Talc	—	50	50
Camel-White	—	60	—
Foral® 85	—	20	80
Amberol ST-149	—	20	—
Resin ST-5115	15	—	—
Titanium Dioxide	—	10	—
Indopol® H-100	150	150	150
Aluminum Paste	—	—	as desired

Concrete Bond Coat Adhesive

Polyvinyl Acetate Emulsion	40
Gelvatol 20-30	55
Dibutyl Phthalate	5

Brick Mastic

Formula No. 1

(Acrylic Resin)

Rhoplex® AC-64	100.0
Triton® X-405	1.4
Tamol® 850	0.7
Foamaster	0.4
Biocide	0.2
Propylene Glycol	5.0
Varsol	2.0
Camel Carb®	50.0
#60 Sand	25.0
#45 Sand	25.0
Sno Cal Clay	20.0
Cellufloc®	6.0
Hi Sil 422	5.0
Acrysol® ASE-60/Water (1/1)	1.5
Water	1.5

Properties:

Pigment/Binder Ratio	2.2/1
Solids	79.4%

No. 2

(Acrylic Resin)

Rhoplex® AC-64	100.0
Triton® X-405	1.8
Tamol® 850	0.8
Nopco® NXZ	0.6
Biocide	0.2
Ethylene Glycol	17.5

Varsol	10.0
Texanol	5.0
Pennsylvania Limestone	300.0
Petinos #45 Sand	150.0
Camel Carb®	75.0
Cellulose Floc #CP-40	4.5
Pigment (umber)	2.0
Acrysol® ASE-60/Water (1/1)	22.0
Water	41.0

Properties:

Pigment/Binder Ratio	8.8/1
Solids	81.7%
Visc. (Brookfield HAT #7/100 rpm)	208 poise
Freeze-Thaw Stability	Pass–5 cycles
Storage Stability, 1 mo. @ 50 C	Pass

Masonry Patching Compound

Formula No. 1

(Epoxy/Resin)

A	**Epotuf 6130**	310
	Glyceryl Monooleate	12
	Camel Carb®	750
B	**Genamide 250**	127
	Thixcin E	5
	Camel Carb®	350

	No. 2	No. 3
	(Epoxy/Resin)	
Polylite Resin 8230	522	522
Thixcin E	20	5
Cobalt Naphthenate (6%)	10	10
Camel Carb®	925	925
Lupersol Delta	8	8

Note:

The catalyst (**Lupersol Delta,** methyl ethyl ketone peroxide) and the accelerator (cobalt naphthenate) called for in the above formulas should never be mixed directly together since a violent reaction may result.

No. 4

(Acrylic Latex)

1.	**Ucar® 153** (55% N.V.)	490.00
2.	**Daxad 30**	5.00
3.	**Triton® 405**	4.50
4.	**Colloid 681F**	2.00
5.	**Benzoflex 9-88**	85.00
6.	Butyl Cellosolve Acetate	8.00
7.	**Atomite®**	150.00
8.	**Snobrite Clay**	150.00
9.	**Carbopol® 934**	2.00
10.	Water	85.00
	AMP-95	5.00
11.	**Keystone #1 Dry**	200.00
12.	Raw Umber*	1.65

---

* Use optional, not required

Procedure:

Charge pony mix chamber or the equivalent with ingredients 1–4. Mix well, then add ingredients 5 and 6. Slowly, add dry ingredients 7 and 8 stepwise, mixing thoroughly with each addition. Continue to mix until pigment is well dispersed (20 min). Then add 9 (**Carbopol® 934**), sifting in slowly until well incorporated. At this point slowly add the preblended water and **AMP-95** solution (step 10). The mixture will increase in viscosity since the **AMP-95** will swell the **Carbopol®.** (Note: **AMP-95** functions as a secondary dispersant whose primary purpose is a pH adjuster. Also, it swells the **Carbopol®** to develop the required nonslump character). Mix until compound is smooth and uniform. Finally, add the silica sand 11 gradually and mix until the mass is homogeneous. Colorant, step 12, is optional, depending on final color desired.

Time total mixing cycle so that final step (silica sand addition) will be as short as possible to minimize any possible abrasion.

Special Note:
Replace 3 gal of water with ethylene glycol if freeze-thaw stability is desired.

No. 5

(Epoxy, Sand-Filled)

A	**Epotuf 37-128**	220.0
	Thixatrol ST	8.0
B	**Epotuf Hardener 37-612**	132.0
	DMP-30	11.0
	Thixatrol ST	5.5
C	**#1/4 Sand**	1280.0

Procedure:
Disperse **Thixatrol ST** in **Epotuf 37-128** on disperser-type equipment. Repeat operation for B. The sand is added prior to use, at which time B is also added to A.

Greenhouse Glaze

(Oil-Based)

Coray 80 (min. oil)	75
Aged Linseed Oil	75
Soya Lecithin	2
Atomite®	450
Putty Filler	323
Mistron Vapor®	75

Note:
The most important characteristics required of a greenhouse glaze are a) very slow drying, b) a high degree of internal softness, c) good adhesion to metal and glass, and d) the ability to be handled and formed for application without sticking to the hands. This formulation has been carefully balanced to achieve these properties.

Glazing Compound
Formula No. 1
(Acrylic Resin)

	lb	gal
Oleic Acid	15.0	2.2
Triethanolamine	8.0	0.8
Rhoplex® LC-40	250.0	28.0
Triton® X-405	9.0	1.0
Paraplex® WP-1	75.0	9.0
Silane Z-60-40	0.6	—
Troysan PMA-30	4.0	0.4
Ethylene Glycol	24.0	2.6
Ti-Pure R-900	10.0	0.3
Atomite®	925.0	41.2
Duramite	325.0	14.5

Procedure:

The triethanolamine and oleic acid are premixed until completely reacted with each other in a chamber of a suitable heavy-duty mixer. Each ingredient thereafter is well mixed before the addition of the next and is added in the order given above. *Precaution:* The **Triton® X-405** which is the latex stabilizer and an emulsifier must be added to the **Rhoplex® LC-40** in the mixing chamber and well incorporated prior to the **Paraplex® WP-1** plasticizer addition. Dry ingredients are added slowly to the mixture and allowed to mix until the pigment is thoroughly dispersed. This is normally about $1/2$ h, longer if necessary, depending upon the efficiency of the mixer. Package the finished compound immediately after the mixing cycle to avoid as much loss of moisture as possible.

No. 2
(Oil-Based)

Degummed Raw Soya	70
Raw Linseed Oil	—
Aged Linseed Oil	20
Blown Soybean Oil (Z-4 visc.)	—
Soya Fatty Acids	2
Mineral Spirits	18
Manganese Naphthenate (6%)	—
Atomite®	320
Putty Filler	555
Mistron Vapor®	15

No. 3
(Oil-Based)

Raw Selectol	85.0
Special Varnish Soya	43.0
Soya Fatty Acids	2.0
Soya Lecithin	0.2
Atomite®	845.0
Rutile Titanium dioxide	10.0
Bentone® 34	11.0
Denatured Alcohol	3.8

Note:

This glazing compound is designed to have the very best appearance and application properties. This is manifested by extreme smoothness, excellent handling characteristics and brilliant whiteness. It is necessary to use only the finest ground calcite pigment, **Atomite®**, in combination with a treated clay (**Bentone® 34**) in order to obtain the desired smooth plastic nature of this compound. This smoothness would be destroyed if coarser grades of ground calcite were used in the formulation.

Hot-Melt Glazing Compound

Thermoplastic Elastomer	3.84
Tackifying Resin	7.69
Amoco® Resin 18-290	9.61
Amoco® H-1500	24.98
Calcium Carbonate	42.27
Silica Extender	9.61
Titanium Dioxide	1.92
Antioxidant	0.04
Stabilizer	0.04

Procedure:

Prepare in a sigma blade mixer equipped with a jacket for steam and cold water using the following steps:

Start	Steam on to 149 C (300 F); charge polybutene, antioxidant, stabilizer, calcium carbonate, silica, TiO_2.
5 min	Add copolymer.
10 min	Add **Resin 18.**
30 min	Add tackifying resin.
65 min	Dump

Knife-Grade Glazing Compound

Raw Soya Oil (A_5 visc.)	2.40
Blown Soya Oil (Z_4 visc.)	5.00
Bodied Linseed (Z_4 visc.)	0.80
Amoco® H-100	4.50
Soya Fatty Acid	0.30
Calcium Carbonate	27.80
Marble Dust	55.70
Talc	3.50

Procedure:

In the laboratory the vehicle ingredients are blended in a heavy-duty sigma blade mixer for 20 min. Filler components are added in the order shown and mixed thoroughly over a 40-min period. Total processing time is approximately 1 h.

Glass Window Sealant

	Formula No. 1	No. 2
	(Polybutene/Resin)	
Butyl Elastomer	15.6	16.6
HAF N330	23.5	25.5
Phenolic Resin	29.5	25.6
Amoco® H-1900	23.5	17.5
Amoco® 18-290	—	5.9
Ethylene Propylene Rubber	7.9	8.9

Coating, Putty, and Adhesive Compound

(Epoxy-Resin)

Epoxy Resin (liquid, unmodified)	75
Pine Oil	20
Phenol (as accelerator)	0–5
Aerosil® 200, 300, or 380	1–4
Diethylene Triamine	10

Note:

Fillers and pigments can be added as desired.

Window Putty

(Styrene-Butadiene/Resin)

Ameripol 1009	50
Ameripol 1006	50
Cumar® MH 2½	95
Calcene TM	160
Heptane	200

Nondeteriorating Construction Putty

	Formula No. 1	No. 2
	(White)	*(Black)*
FA	100.0	100.0
Whiting	285.0	285.0
Titanium Dioxide	15.0	—
SRF Black	—	15.0
MBTS	0.4	0.4
DPG	0.1	0.1
Benzothiazyl Disulfide	2.0	2.0

Commerical Grade Putty

(Oil-Based)

Coray 60 (min. oil)	80
Aged Linseed Oil	20
Soya Fatty Acids	2
Atomite®	300
Putty Filler	588
Mistron Vapor®	10

Note:

Commercial putty is designed to have superlative knifing, handling, can stability, and brilliant whiteness at highly competitive raw material costs. In order for this compound to perform very satisfactorily, however, most surfaces should be primed before using this material because of the strong tendency of the mineral oil portion to migrate into substrates.

White Linseed Oil Putty
(With and Without White Lead)

	Formula No. 1	No. 2
	(Oil-Based)	
Raw Linseed Oil	80	80
Aged Linseed Oil	20	20
Soya Fatty Acids	1	1
Atomite®	320	200
Putty Filler	569	589
Mistron Vapor®	10	10
B.C. White Lead	—	100

Metal Sash Putty

(Oil-Based)

Raw Linseed Oil	70.0
Aged Linseed Oil	20.0
Soya Fatty Acids	2.0
Mineral Spirits	18.0
Manganese Naphthenate (6%)	0.5
Atomite®	320.0
Putty Filler	555.0
Mistron Vapor®	15.0

Stainless Putty

(Oil-Based)

Raw Linseed Oil	70.0
Blown Soybean Oil (Z-4 visc.)	20.0
Soya Fatty Acids	2.0
Mineral Spirits	18.0
Manganese Naphthenate (6%)	0.5
Atomite®	320.0
Putty Filler	555.0
Mistron Vapor®	15.0

Plumber's Putty

(Oil-Based)

#5 Blown Selectol	37
Coray 60 (min. oil)	111
Atomite®	150
Snowflake White	668
Mistron Vapor®	20
Bentone® 34	14

Note:

Plumber's putty is essentially a bedding compound which is used as a seal in installing various plumbing fixtures. It is important that it display excellent can stability and handling characteristics. Also, it is important that it does not penetrate readily into the ceramic parts or into other substrates and that it remain relatively soft after being applied. For best results the compound suggested above should be prepared by the ingredients being added in the order given. The **Bentone® 34** should be withheld until all other ingredients are thoroughly mixed. In this way the **Bentone® 34,** which produces a "gelled effect," has a maximum influence on reducing flow, creating smoothness and developing a much desired plasticity to the putty.

Clear Weatherable Sealant

(Rubber/Resin)

Kraton® G 1652	67.0
Kraton® GX 1701	33.0
Arkon® P-85	167.0
Kristalex® 1120	67.0
Silane A-189	2.3
Irganox® 1010	1.7
Tinuvin® 327	0.7
Tinuvin® 770	0.7
Tolusol 25	96.0
Propyl Acetate	24

Properties:

Hardness, Shore A	26
180° Peel strength against glass, original*	40 pli
180° Peel strength against glass,	
after 1 week's soaking in water	40 pli
Color change or surface cracks	
after 1000 h in carbon arc weatherometer	None
Service temp. range**	—15–70 C

---

* 0.1 in. thick, cast on canvas against glass, pulled at 2 in./min.

** Can be bent at –15 C without cracking. No slump after one week on vertical surface at 70 C.

Gunnable Dispersion-Based Sealants

Formula No. 1

(Acrylic)

Acronal 81 D (pH 8)	329.0
Plastilit White Paste*	100.0
Plastilit 3060	40.0
Emulan® OG	1.0
Pigment Disperser N	0.5
Nopco® NXZ	0.5
Silica HDK 20	9.0
Microtalk AT 1	80.0
Omya® BLP 3	440.0

Grind on 3-roll mill.

---

*Plastilit White Paste:

Plastilit 3060	42
Titanium White RN 56	58

No. 2

(Acrylic)

Acronal 81 D (pH 8)	329.0
Oppanol B 3 White Paste*	140.0
Emulan® OG	1.0
Pigment Disperser N	0.5
Nopco® NXZ	0.5
Microtalk AT 1	89.0
Omya® BLP 3	440.0

---

* **Oppanol B 3** White Paste:	
Oppanol B 3	58.5
Titanium White RN 56	41.5

No. 3

(Acrylic)

Acronal 81 D (pH 8)	320.0
Plastilit 3060	75.0
Plastilit Grey Paste*	10.0
Emulan® OG	1.0
Pigment Disperser N	0.5
Nopco® NXZ	0.5
Silica HDK N 20	8.0
Microdol 1	140.0
Omya® BLP 3	445.0

---

* **Plastilit Grey Paste:**	
Plastilit 3060	50
Iron Oxide Black	50

No. 4

(Acrylic)

Acronal 81 D (pH 8)	400
Plastilit 3060	70
Plastilit Grey Paste	10
Emulan® OG	2

Pigment Disperser N	1
Nopco® NXZ	1
Silica HDK N 20	10
Microdol 1	50
Microtalk AT 1	106
Omya® BLP 3	350

No. 5

(Acrylic)

Acronal 355 D (pH 8)	275.0
Plastilit 3060	127.0
Plastilit Grey Paste	14.0
Pigment Disperser N	1.5
Nopco® NXZ	1.5
Emulan® OG	3.0
Rewopol® NOS 25	4.0
Ethanediol	7.0
Silica HDK N 20	6.0
Omya® BLP 3	561.0

Note:

The adhesion of dispersion-based sealants to absorbent substrates is usually sufficient even without a primer coat. However, in order to be on the safe side, especially for binding dust particles, it is advisable to apply a primer coat of diluted sealant (e.g., 1 part sealant, 3 parts water) to the flanks of the gap.

The sealant is applied into the gaps by a hand or pneumatically operated gun. The surface of the joint is subsequently smoothed with a wet flat brush.

Exterior gaps should not be filled in rainy weather or at temperature below +5 C because immediately after it has been applied into the gap, a dispersion-based sealant is still sensitive to water and requires 30–60 min for forming an adequately thick skin, depending on the temperature and relative humidity. The equipment for applying the sealants should be cleaned with water immediately after use because dried residues can be removed only mechanically.

Uses:

Sealants based on **Acronal** can be applied on all absorbent substrates, such as concrete, aerated concrete, asbestos cement, plaster, gypsum, and wood. They can be used for sealing all types of interior and exterior joints and expansion joints (interior and exterior with an expansion of 10–15%. The joints must be designed according to DIN 18540.

Dispersion-based sealants are unsuitable for sealing joints which are steadily exposed to water or moisture, or which are subject to large movements caused by expansion and shrinkage.

Gunnable Caulks

Formula No. 1

1. **Ucar® 153** (55% N.V.)	285.00
2. **Triton® X-405**	9.15
3. **Composition T**	10.50
4. **Benzoflex 9-88**	141.00
5. Ethylene Glycol	9.00
6. Mineral Spirits	33.00
7. **Atomite®**	650.00
8. **Ti-Pure® R-901**	10.00
Reduction	
9. **Ucar® 153** (55% N.V.)	145.00

Procedure:

Charge the chamber of a suitable mixer (e.g., Baker Perkins double sigma-blade type or its equivalent, that is, one which will not incorporate air) with about two-thirds of the total acrylic latex binder content (step 1). This amount is required in order to develop the desired mixing consistency and simultaneously to enhance pigment dispersion in a very reasonable length of time. Each ingredient is added in the order given above and well mixed prior to the addition of the next ingredient.

Note: For best results the **Triton® X-405**, a latex stabilizer and an emulsifier, should be added prior to the plasticizer addition (i.e., **Benzoflex 9-88**). The **Composition T**, a pigment disperser, is added to the mix (step 3) and allowed to be dissolved as thoroughly as possible in the **Ucar® 153** binder in order to effectively disperse the pigment portion. The mixing cycle is then continued in a closed chamber to avoid solvent evaporation until such time as the pigment is completely dispersed in the media. This is normally about $1/2$ h, longer if necessary, depending upon the efficiency of the mixer. At this point the caulk is reduced slowly with the remaining binder portion (step 9). Mixing is continued until the mixture is completely homogeneous in texture.   Package the finished caulk immediately after the end of the mixing cycle.

No. 2

(Acrylic Latex)

Ucar® 153 (55.5% T.S.)	442.42
Triton® X-405	9.73
Composition T	10.86
Flexol® TCP	60.13
Aroclor 1254 or 6062	67.49
Varsol No. 1	27.64
Atomite®	722.30
Ti-Pure® R-901	8.36

Procedure:

Add in order shown. Formulation is mixed 2 h in a 1-gal Baker-Perkins sigma-blade type mixer at 79 rpm. Enclosed mixer minimizes moisture loss.

No. 3

(Polyester/Resin)

1. Rhoplex® LC-40	215.09
2. Triton® X-405	9.46
3. Composition T	10.65
4. Paraplex® WP-1	124.21
5. Esso Varsol #1	26.91
7. Atomite®	692.06
8. TiPure R-901	17.72
Reduction:	
9. Rhoplex® LC-40	215.08

Procedure:

It is important in the preparation of this caulk: a) that one-half of the Rhoplex® LC-40 binder be charged into the mixing chamber of a

Baker Perkins double sigma-blade type mixer (or any other good mixer which does not incorporate air) followed by the other ingredients in the order given if best results are to be obtained, and b) that each ingredient is well mixed before the addition of the next. (*Precautions:* All of the **Triton® X-405** which is the latex stabilizer and an emulsifier must be added to the step one **Rhoplex® LC-40** and well incorporated prior to the **Paraplex® WP-1** plasticizer addition. Of equal importance is that the **Composition T**, pigment disperser, be as thoroughly dissolved as possible in the step one **Rhoplex® LC-40** to be of maximum effectiveness when the pigment portion is introduced), c) that the mixing be continued for about $1/2$ h with a cover over the top of the mixing chamber in order to avoid loss of solvent and to thoroughly disperse the pigments, d) the other half of the **Rhoplex LC-40** binder should then be added slowly to the compound and mixing continued until the caulk is homogeneous, and e) that the product be packaged immediately after the mixing cycle is completed.

No. 4

(Polyester/Resin)

Experimental Resin ZR-172	2336.00
Camel Carb®	1800.00
Texas Talc # 2619	408.00
Ti-Pure® R-901	96.00
Thixatrol ST	194.00
Paraplex® G-50	388.00
Monoplex DBS	19.40
Cobalt Naphthenate (6%)	2.40
Zinc Naphthenate (8%)	12.10
Silane A-174	4.85
Xylene	116.00

Procedure:

Prepare under nitrogen. Charge **Camel Carb®, Texas Talc, Thixatrol ST, TiPure R-901** and mix for several minutes. Charge **Experimental Resin ZR-172** and mix for 10 min and then add **Paraplex® G-50** and **Monoplex DBS** to sealant and mix for 40 min. Charge driers, and mix for few minutes. Premix silane and xylene, and then add the mixture to sealant; mix for 10 min.

No. 5

(Polyester/Resin)

Experimental Resin ZR-172	2336.00
Camel Carb®	408.00
Ti-Pure® R-901	96.00
Thixatrol ST	194.00
Paraplex® G-50	388.00
Monoplex DBS	19.40
Cobalt Naphthenate (6%)	2.40
Zinc Naphthenate (8%)	12.10
Silane A-174	4.85
Xylene	116.00

Procedure:

Prepare under nitrogen. Charge **Camel Carb®, Thixatrol ST, Ti-Pure® R-901** and mix for several minutes. Charge **Experimental Resin ZR-172** and mix for 10 min and then add **Paraplex® G-50** and **Monoplex DBS** to sealant and mix for 40 min. Charge driers, and mix for few minutes. Premix silane and xylene, and then add the mixture to sealant, mix for 10 min.

No. 6

(Polyester/Resin)

Experimental Resin ZR-172	2336.00
Camel Carb®	1800.00
Texas Talc #2619	408.00
Ti-Pure® R-901	96.0
Thixatrol ST	194.00
Cobalt Naphthenate (6%)	2.40
Zinc Naphthenate (8%)	12.10
Silane A-174	4.85
Xylene	216.00

Procedure:

Prepare under nitrogen. Charge **Camel Carb®, Texas Talc, Thixatrol ST, Ti-Pure® R-901** and mix for several minutes. Charge **Experimental Resin ZR-172**, and mix for 50 min. Charge driers, and mix for few minutes. Premix silane and xylene, and then add the mixture to sealant, mix for 10 min.

No. 7

(Polybutene)

Polybutene H-1900	8.30
Butyl Sol'n.	7.69
Hydrous Silicate	3.14
Denatured Alcohol	0.16
Hydrogenated Rosin Ester	0.61
Calcium Carbonate	54.44
Polybutene L-14	10.70
Talc	7.25
Aluminum Paste	2.54
Mineral Spirits	5.17

Procedure:

In a sigma-blade mixer use the following steps: **Polybutene H-1900** and butyl solution are added and mixed thoroughly (10–15 min). Alcohol and the thixotropes are then added and mixed until swelling and homogeneity of the mix is obtained (15–20 min). $CaCO_3$ is added slowly; if mixing becomes labored, add portions of **Polybutene L-14**. Mix to a smooth consistency (15–20 min). Add talc and remainder of **L-14**; mix until homogeneous (15–20 min). Add mineral spirits and aluminum paste; mix until thoroughly dispersed (5–10 min).

	No. 8	No. 9
	(Polybutene/Butyl Rubber)	
1. **AD-50** (50% N.V., Rule 66)	247.0	247.0
2. **Indopol® H-100** (or equivalent)	168.0	168.0
3. **Castung 403 Z-3**	32.0	32.0
4. Cobalt Naphthenate (6%)	0.4	0.4
5. **Atomite®**	350.0	300.0
6. **Mistron Vapor®**	300.0	300.0
7. **Ti-Pure® R-900**	15.00	—
8. Aluminum Paste (73.5% N.V.)	—	35.0
9. Mineral Spirits (Rule 66)	59.0	60.0

Procedure:

A heavy-duty double sigma-blade Baker-Perkins mixer is the type of equipment recommended for making these formulations and the suggested

technique is as follows. Roughly one-half of the butyl cut is added to the mixer, followed by the **Atomite®**, rutile titanium dioxide (for white caulk) and **Mistron Vapor®**, in that order. The **Indopol® H-100** and/or a sufficient amount of the remaining butyl cut should be added as necessary to incorporate the pigments rapidly and yet allow a stiff enough mixing consistency for efficient incorporation of the pigments. When the compound has become homogeneous, the remainder of the butyl cut (if any) and the dehydrated castor oil, **Castung 403Z-3**, is added and allowed to work in, followed by the cobalt drier mixed in the solvent and added in small increments. Formula No. 9 retains a shiny aluminum look if the aluminum paste is withheld until near the end of the mixing cycle and given only sufficient mixing time to insure its uniform distribution.
Note:

Nonstaining grades of butyl rubber may offer some advantages in color after long exposure and this should be evaluated where maximum color retention is considered to be of major importance.

No. 10

(Oil-Based, Polybutene)

Blown Soybean Oil (Z-4 visc.)	196
Polybutene (Z-6–Z-7 visc.)	88
Soya Fatty Acids	16
Atomite®	950
Thixcin R	23
Cobalt Naphthenate (6%)	4
Mineral Spirits	107

Properties:

PVC	54.2%
Nonvolatile by volume	83.1%

No. 11

(Oil-Based, Polybutene)

Blown Soybean Oil (Z-4 visc.)	245
Polybutene (Z-6 to Z-7 visc.)	110
Soya Fatty Acids	16
Atomite®	900

Mistron Vapor®	100
Cobalt Naphthenate (6%)	4
Mineral Spirits	53

Properties:

PVC	49.4%
Nonvolatile by volume	91.9%

	No. 12	No. 13
	High Quality	Low Cost
	(Oil-Based, Polybutene)	
Blown Soybean Oil (Z-4 visc.)	245	163
Polybutene (Z-6 to Z-7 visc.)	110	73
Soya Fatty Acids	16	16
Soya Lecithin	—	4
Atomite®	800	1000
International Fiber 1	200	—
Cobalt Naphthenate (6%)	4	2
Mineral Spirits	55	39
Methocel® (3% sol'n. in water)	—	148

Properties:

PVC (%)	49.4	59.7
Nonvolatile by volume (%)	91.5	77.5

Notes:

For a brilliant white use 10 lb of rutile titanium per 100 gal formulation. For an aluminum color use 20 lb of **Alcoa's 231** aluminum paste or equivalent and 10 lb of **Alcoa S-242** desiccant or equivalent per 100 gal formulation. Aluminum should *not* be used in the low-cost formulation since the water in this formulation will cause gassing.

The high-quality formulation shown above has been compounded to give excellent ease of gunning and yet have good nonslump under normal conditions. However, this material may be considered too firm when used in winter weather or cold climates and conversely too soft and not sufficiently slump-resistant in very hot summer days or in tropical climates. The formulation can be adjusted for these conditions by removing some of the **International Fiber** and replacing it with an equal weight of **Atomite®** for softer compounds and reversing this procedure for firmer, more slump resistant compounds.

The low-cost formulation will become very stiff at temperature below the freezing point of water due to the freezing of the internal aqueous emulsion. This can be prevented by using a colloidal gel solution which was prepared with water containing 5% by volume of ethylene glycol. The caulk thus modified handles satisfactorily over a wide temperature range.

	No. 14	No. 15
	(Polybutene/Resin)	
EX-214	100.0	—
EX-214 (pelletized, preswollen with VM&P Naphtha 1:1 ratio)	—	200.0
Indopol® H-100	125.0	125.0
Microwhite 95	500.0	500.0
Asbestine 3X	250.0	250.0
Ti-Pure® R-900	12.5	12.5
Super Hi Flash Naphtha	150.0	—
VM&P Naphtha	—	50
Betaprene® B-140	15	—

Procedure (No. 13):

Time

0 min   Cold water on, charge mixer with all **Asbestine**, TiO_2, **Indopol® H-100**, **Betaprene®**, and $1/4$ **Microwhite**.

1 min   Add polymer.

10 min   Slowly add remaining **Microwhite**.

25 min   Add solvent slowly.

60 min   Dump.

Procedure (No. 14):

Time

0 min   Cold water on, charge mixer with preswollen **EX-214** in solvent, all **Asbestine**, TiO_2, $1/2$ **Indopol® H-100**, **Betaprene®**, and $1/2$ **Microwhite**.

10 min   Add remaining **Microwhite**.

15 min   Add remaining **Indopol® H-100**.

25 min   Add remaining solvent.

30–45 min   Dump.

No. 16

(Styrene Butadiene)

Atomite®	475
Rutile Titanium Dioxide	20
Kaydol Mineral Oil (extra heavy)	100
Ameripol 4503 (SBR Crumb)	150
Antioxidant 330	2
Mineral Spirits (regular)	309

Procedure:

Charge a high-shear mixer with **Ameripol 4503, Kaydol** mineral oil, mineral spirits, and antioxidant. This is approximately 80% of total formula volume. Mix until rubber is penetrated by solvent and mixture thickens. Continue mixing until rubber is completely swollen and dispersed. **Atomite®** and titanium dioxide are then added and mixing continued until compound is uniformly smooth. Package immediately.

Note:

Some equipment limitations may require preparation of binder mix by roller milling to insure complete dispersion of the SBR rubber crumbs.

No. 17

(Styrene Butadiene/Resin)

Ameripol 1006	25
Ameripol 1009	75
AgeRite Superlite	3
Primol	35
Atomite®	150
Calcene	75
Asbestos Fines 7T-F1	5
Staybelite® Ester 10	20
Titanium Dioxide	5
Toluene	75
Xylene	50
Heptane	75

Gun and Squeeze-Tube Latex Caulk

(Vinyl Acetate Ethylene)

Amsco Res 7676	404.0
Triton® X-405	11.0
Shancosperse	5.0
Deefo 278	4.0
Benzoflex 9-88	110.0
Tamol® 850	1.2
Mineral Spirits	27.5
Ethylene Glycol	21.0
Titanium Dioxide A-410	5.0
Camel-White	650.0

Properties:
Wt./gal   12.3 lb

Butyl Caulk

Formula No. 1

(Polybutene)

Amoco® Polybutene H-1900	14.00
Butyl Solution	21.48
Blown Soya Oil (Z2 visc.)	2.80
Calcium Carbonate	37.35
Talc	14.01
Titanium-Calcium Pigment	4.67
Cobalt Drier (6%)	0.03
Mineral Spirits	5.66

Procedure:

The caulk is mixed in a sigma-blade mixer. **Amoco® Polybutene,** butyl rubber solution, and drying oil are added and mixed well (5–10 min); then the fillers are slowly added to the mixing vehicle in the order shown above. After the addition of titanium–calcium pigment, the mass is mixed to a smooth paste (about 35 min). The drier is added along with sufficient mineral spirits to obtain the desired consistency. Mixing is continued until the mass is uniform (about 3–5 min).

	No. 2	No. 3	No. 4
	(Polybutene)		
Amoco® H-100 Polybutene	20.00	21.00	18.00
Amoco® H-300 Polybutene	—	—	1.00
Keltrol	4.39	4.39	4.39
Tall Oil Fatty Acid	0.49	0.49	0.49
Calcium Carbonate	50.80	50.50	51.30
Platelet Talc	5.80	5.40	9.40
Diatomaceous Silica	3.20	3.20	3.20
Hard Clay	11.70	11.40	8.60
Mineral Spirits	3.33	3.33	3.33
Cobalt Drier (6%)	0.29	0.29	0.29

Procedure:

The vehicle ingredients blend in a dual-arm sigma-blade mixer. Fillers are added incrementally in the order listed and then mixed 30 min. The mixture is thinned to desired viscosity with mineral spirits, the driers added, and the caulk mixed to a smooth consistency.

No. 5

(Oil-Based, Polybutene)

Blown Soybean Oil (Z-4 visc.)	196
Polybutene (Z6 to Z7 visc.)	88
Soya Fatty Acids	16
Calcium Carbonate	950
Thixcin 25C	92
Mineral Spirits	38
Cobalt Naphthenate (6%)	4

No. 6

(Oil-Based, Polybutene)

1. Blown Soybean Oil (Z-4 visc.)	49
2. **Indopol® H-1500 or Polyvis 150-SH**	226
3. Soya Fatty Acids	16
4. **Atomite®**	950
5. **Thixcin R**	23
6. Mineral Spirits	110

Properties:

PVC	54.0%
Nonvolatile by volume	83.1%

Procedure:

Combine ingredients 1, 2, and 3 above in a suitable mixer (e.g., sigma double-blade Baker Perkins) and let mix until well blended. Then add **Atomite®** as rapidly as is consistent with efficient mixing. Add the **Thixcin** and continue mixing for 1 h. The mineral spirits is added as the last ingredient and mixed until uniformly smooth and then canned.

Note:

If a slightly firmer type of caulk behavior can be tolerated along with a slight tendency to skin at the exposed edges after prolonged exposure, it is possible to obtain substantially complete nonpenetration character by increasing the amount of soybean oil and simultaneously lowering the polybutene.

No. 7

(Polybutene/Butyl Rubber)

Bucar 5214	120
Indopol® H-100	155
Super Sta-Tac® 80	16
Soya Fatty Acids (RO-11S)	4
Atomite®	600
Asbestine 3X	190
Titanium Dioxide	14
Mineral Spirits	176

Procedure:

A heavy-duty, double sigma-blade Baker Perkins mixer or equivalent is recommended, utilizing the following procedure: Pre-swell the **Bucar 5214** with an equal weight of mineral spirits. Drum roller operation for 2–3 h is suggested to ensure uniform absorption of mineral spirits on butyl pellet surfaces. At this point, the **Bucar 5214** is not yet completely swollen, and unless a very heavy-duty caulk mixer having close bladewall clearances is available, further soak time of **Bucar 5214** in solvent, up to 72 h, is advised.

Charge the mixer with the butyl cut, **Super Sta-Tac® 80** (precut in part of the mineral spirits content) and soya fatty acids. While mixing, add

Atomite®, **Asbestine 3X**, and titanium dioxide, alternating the pigment charge with enough polybutene to incorporate the total pigment load.

The balance of polybutene is added after the pigment is well dispersed. Lastly, add the remaining mineral spirits as rapidly as is consistent with efficient mixing. Continue to mix in a closed chamber until the compound is smooth and homogeneous ($\approx$ 1 h).

Properties:	Spec Limits	Results
Bubble Formation	25% max.	pass
Tenacity	no cracking or separation	pass
Shrinkage	25% max.	22.1%
Slump	0.15 in. max.	none
Extrudability	9.0 s/ml max.	7.13 s/ml
Stain	2.5 stain index mas.	pass
Tack-Free Time	within 24 h	pass

Adhesion loss, cracking, discoloration:

Discoloration		none
Adhesion Loss	#2 Rating max.	0
Center Crack	#2 Rating max.	0
Edge Crack	#2 Rating max.	0
Bond Cohesion	1.5 in.^{2}	none

(Glass, Aluminum, Mortar)

No. 8

(Polybutene/Butyl Rubber)

Bucar 5214	120
Indopol® H-100	155
Super Sta-Tac® 80	16
Soya Fatty Acids (**RO-11S**)	4
Atomite®	710
Mistron Vapor®	82
Titanium Dioxide	14
Mineral Spirits	176

Procedure:

A heavy-duty, double sigma-blade Baker Perkins mixer or equivalent is recommended, utilizing the following procedure: Pre-swell the **Bucar**

5214 with an equal weight of mineral spirits. Drum roller operation for 2–3 h is suggested to ensure uniform absorption of mineral spirits on butyl pellet surfaces. At this point, the **Bucar 5214** is not yet completely swollen, and unless a very heavy-duty caulk mixer having close bladewall clearances is available, further soak time of **Bucar 5214** in solvent, up to 72 h, is advised.

Charge the mixer with the butyl cut, **Super Sta-Tac® 80** (precut in part of the mineral spirits content) and soya fatty acids. While mixing, add **Atomite®**, **Asbestine 3X**, and titanium dioxide, alternating the pigment charge with enough polybutene to incorporate the total pigment load.

The balance of polybutene is added after the pigment is well dispersed. Lastly, add the remaining mineral spirits as rapidly as is consistent with efficient mixing. Continue to mix in a closed chamber until the compound is smooth and homogeneous (≈ 1 h).

Caulk

Formula No. 1

(Solvent/Polybutene/Butyl Rubber)

Bucar 5214	8.6
Super Sta-Tac® 80	1.7
Indopol® H-100	10.8
Camel Carb®	43.2
IT-3X	21.6
Titanium Dioxide	1.0
VM&P Naphtha	12.9

Procedure:

Swell rubber in solvent. Charge to a sigma-blade mixer (or suitable mixing equipment). Charge all resin, talc, titanium dioxide, $1/2$ polybutene, and $1/2$ calcium carbonate. Mix. Add remaining calcium carbonate. Add remaining polybutene. Mix until system is homogeneous.

Properties:
Solids	87%

No. 2
(Architectural, Polybutene/Butyl Rubber)

Polysar XL-50	100.0
Amoco® H-100	200.0
Keltrol® 1001	50.0
Thixatrol GST	6.6
Atomite®	653.0
Asbestine 3X	417.0
Ti-Pure® R-900	67.0
Mineral Spirits	166.0
Cobalt Drier (6%)	0.5
AgeRite Stalite	2.0
Super Beckacite® 2000 (70% in xylene)	5.0

Procedure:

All materials are mixed in a laboratory scale heavy-duty sigma-blade mixer equipped with a compression ram. The mixer is jacketed to allow for water cooling and steam heating.

Time	
0 min	Steam on to approximately 250 F (121 C), add **Thixatrol GST.**
1 min	Add **Amoco® H-100, AgeRite Stalite,** and **Super Beckacite® 2000.**
5 min	Steam off, incrementally add fillers and pigment.
15 min	Cold water on, add **Polysar XL-50,** lower ram.
30 min	Raise ram, add **Keltrol® 1001** incrementally.
35 min	Add mineral spirits and cobalt drier over a 15-min period.
50 min	Mix for 5 min.
55 min	Dump.

No. 3
(Polybutene/Resin)

Kalar 5214	100
Indopol® H-100	125
Microwhite 95	500
IT-3X	250
Titanium Dioxide	12
Super Sta-Tac® 80	20
VM&P Naphtha or Mineral Spirits	150

No. 4

(Polybutene/Butyl Polymer)

EX-245 Pre-Solvated Mixture*	230
Indopol® H-100	70
Indopol® H-1900	25
Atomite®	500
Marble Dust	125
Asbestine 3X	125
Super Sta-Tac® 80	20
Titanium Dioxide	12
Stearic Acid	2
Mineral Spirits	75

* Presolvated Mixture (24 h without agitation)

EX-245	100
Mineral Spirits	100
Indopol® H-100	30

	No. 5	No. 6	No. 7	No. 8
	(Solvent-Based, Butyl Polymer)			
EX-245	100	100	100	100
Mineral Spirits	130	130	150	100
Indopol® H-100	—	30	30	—

Procedure:

In order to produce a lump-free, homogeneous mixture of **EX-245**, solvent, and polybutenes, the following procedures are applicable for both laboratory and production quantities. Weigh out the correct proportions of **EX-245**, solvent, and polybutene. In a suitable container, blend the polybutene and solvent to insure complete dissolution. Add the pelletized **EX-245** to the container containing the solvent, polybutene mixture. (*Note:* The amount of solvent mixture added should completely cover the pellets of **EX-245**.) Let stand without any agitation for at least 24 h.

If the proportions of **EX-245**, solvent, and polybutene are correct, after 24 h, a highly gelled, semiuniform mass is obtained. In some instances, peculiar to the proportions present, some supernatant liquid will be present. This total mixture can then be charged to the mixer and worked for 1–5 min to produce a uniform and homogeneous blend of the starting materials. A lump-free, creamy **EX-245**-based intermediate is obtained. After homogenization is complete, fillers and other ingredients can be added in the normal fashion.

	No. 9	No. 10
	(Acrylic Resin)	
Rhoplex® LC-45 (wet)	100.0	100.0
Triton® X-405	1.6	1.6
Water	5.2	5.2
Composition T (Calgon T)	1.8	1.8
Plasticizer	32.5	32.5
Varsol #1	7.2	7.2
Ethylene Glycol	—	3.5
Tamol® 850	0.4	0.4
Camel-Tex	190.1	190.1
Ti-Pure® R-901	4.9	4.9
Properties:		
Pigment/Binder Ratio	3/1	3/1
% Solids (wt.)	86.0	86.1

No. 11

(Acrylic Resin)

1.	Oleic Acid	21.0
2.	Triethanolamine	12.0
3.	**Natrosol® 250 HR** (3%)	31.0
4.	**Rhoplex® LC-40** (55)	227.0
5.	**Triton® 405**	9.0
6.	**Paraplex® WP-1**	75.0
7.	**Silane Z-6040**	0.6
8.	**Ti-Pure® R-900**	25.0
9.	**Atomite**	750.0

Reduction:

10.	**Rhoplex® LC-40** (55)	200.0

Procedure:

The triethanolamine and oleic acid are premixed until completely reacted with each other in a chamber of a suitable heavy-duty mixer. Each ingredient thereafter is well mixed before the addition of the next and is introduced into the mixing chamber in the order given (i.e., 3 through 9). (*Precaution:* All of the **Triton® X-405** which is the latex stabilizer and an emulsifier must be added to the initial (step 4) **Rhoplex® LC-40** and well incorporated prior to the **Paraplex® WP-1** plasticizer addition.) The **Atomite®** and rutile pigment are added slowly to the mixture and allowed to mix until the pigment is thoroughly dispersed. This is normally about $1/2$ h, longer if necessary, depending on the efficiency of the mixer. The remaining **Rhoplex® LC-40** binder (step 10) is added slowly to the mix as the reduction vehicle. Continue to mix until the caulk is homogeneous. Package immediately to avoid excessive loss of moisture.

No. 12

(Acrylic Latex)

Ucar® 153	400
Triton® X-405	5
Composition T	7
Benzoflex 9-88	75
Mineral Spirits	20
Ethylene Glycol	10
Atomite®	300
Ti-Pure® R-901	8
Extendospheres XL-100	50

Properties:

Wt./gal	8.75 lb
Total Solids	77.0%
Filler/Latex Solids	1.6/1

No. 13

(Acrylic Latex)

Ucar® 153 (55.5% T.S.)	433.70
Triton® X-405	9.75
Composition T	10.65
Flexol® TCP	58.96
Paraplex® WP-1	66.13
Varsol No. 1	27.12
Atomite®	708.06
Ti-Pure® R-901	8.36

Procedure:

Add in order shown. Formulation is mixed 2 h in a 1-gal Baker-Perkins sigma blade-type mixer at 79 rpm. Enclosed mixer minimizes moisture loss.

	No. 14	No. 15
	Clear	
	(Acrylic Latex)	
Hycar® 4051CG	100	100
Sucrose Benzoate	60	100
Cab-O-Sil® M5	8	—
Acryloid B82	40	100
Di-2-Ethylhexyl Phthalate	110	100
Toluene	55	100
Ethyl Acetate	20	—

No. 16

(Rubber/Resin)

Kraton® GX1657	100.0
Hercules® RES D-151	122.0
Hercules® RES D-44	109.0
Butyl Zimate®	0.3
Irganox® 1010	0.3

No. 17

(Ethylene Vinyl Acetate)

Aircoflex 510	650
Colloids 677	4
Preservative	—
Benzoflex 9-88	50
Strodex PK-90	3
Ti-Pure® R-911	25
Duramite	600
Attagel® 40	20
Talc	50

No. 18

(Vinyl Ethylene)

Amsco Res-7600	650
Nilofoam #7	4
Preservative	—
Santicizer 160	50
Strodex PK-90	3
Ti-Pure® R-911	25
Duramite	600
Attagel® 40	20
Talc	10

	No. 19	No. 20	No. 21
	(Flexible, Polyether/Polypropylene)		
A* Multrathane F-84	223.0	184.0	146.0
Atomite®	223.0	184.0	146.0
B Niax Triol LHT-34	362.0	325.0	275.0
Tributyl Phosphate	8.0	16.0	32.0
Dibutyl Tin Dilaurate	0.5	0.5	0.5
Thixcin R	35.0	25.0	15.0
Atomite®	377.0	616.0	854.0
Ti-Pure® FF	50.0	50.0	50.0

* Tinting A very slightly with a blue will enable the mixing of the two halves to be readily checked visually.

No. 22

(Urethane)

A	**Multrathane F-84**	17.25
	Atomite®	17.25
B	**Niax Triol LHT-35**	28.50
	Tributyl Phosphate	0.66
	Dibutyl Tin Dilaurate	0.04
	Thixcin R	2.74
	Atomite®	29.65
	Ti-Pure® FF	3.91

No. 23

(Polyester/Resin)

Rhoplex® LC-40	430.17
Triton® X-405	9.46
Calgon T	10.65
Paraplex® WP-1	124.21
Silane Z-6040	0.59
Esso Varsol #1	26.91
Tamol® 850	1.27
Atomite®	692.06
Ti-Pure® R-901	17.72

Procedure:

Add ingredients in above order. Add 0.14% ammonium hydroxide (28%) on total formulation weight.

pH of completed caulk should be 7.4–7.5.

Noncuring Caulks

	Formula No. 1	No. 2	No. 3
	(Styrene Butadiene/Resin)		
Solprene 303	100	100.00	100
Piccotex® 100		50.00	—
Picco® 6140-3	100	—	75
Snobrite	—	150.00	—

Atomite®	150	—	100
Hi-Sil 233	10	—	—
Hercolyn® D	10	10.00	10
Wingstay T	2	2.00	2
Titanium Dioxide	—	10.00	—
Ultramarine Blue	—	0.10	—
Mineral Spirits	—	65.00	45
Toluene	75	—	—
Zinc Oxide	10	—	20

Knife-Grade Caulks

Formula No. 1

(Polybutene)

Polybutene H-100	8.23
Blown Soya Oil (Z_4 visc.)	16.16
Tall Oil Fatty Acids	0.49
Calcium Carbonate	47.15
Raw Soya Oil (A_5 visc.)	2.40
Blown Soya Oil (Z_4 visc.)	5.00
Bodied Linseed (Z_4 visc.)	0.80
Polybutene H-100	4.50
Soya Fatty Acid	0.30
Calcium Carbonate	27.80
Marble Dust	55.70
Talc	3.50

Procedure:

Vehicle ingredients are blended in a heavy-duty sigma-blade mixer for 20 min. Filler components are added in the order shown and mixed thoroughly over a 40-min period. Total processing time is approximately 1 h.

Filler:

Talc	23.75
Mineral Spirits	3.93
Cobalt Drier (6%)	0.29

Mix as above.

No. 2

(Oil-Based)

Selectol (raw)	75
Linseed Oil (aged)	75
Atomite®	375
Putty Filler	375
Mistron Vapor®	100

Rope Caulk

(Polybutene)

Indopol® H-1500	311.0
Petrolatum	38.0
Soya Fatty Acids	6.4
Mistron Vapor®	415.0
Snowflake White	375.0
Processed Mineral Fiber	400.0
Rutile Titanium Dioxide	20.0

Note:

Rope caulk is a sealant which is designed primarily as an all-purpose weatherstripping for use around windows and door frames, etc. It can be prepared and extruded *only* in heavy-duty type of equipment. The selection of the ingredients and the ratio in which they are used have to be carefully made to achieve the necessary properties. Rope caulk must be nondrying, have good adhesion to surfaces, and also have sufficient strength after extruding so that it will maintain its rope form even after much handling. The above formulation gives an excellent balance of these desirable characteristics when properly processed. For best results the ingredients should be added in the order given and the mixing continued for some time after the pigments have been added so as to insure complete and uniform incorporation.

Cold Solder

Formula No. 1

(Epoxy)

Epotuf 6130	310
Glyceryl Monooleate	12
Drikalite-AC	750
Genamide 250	127
Thixcin E	5
Drikalite-AC	350

No. 2

(Epoxy)

Epotuf 6130	310
Glyceryl Monooleate	12
Camel-Carb®	750
Genamide 250	127
Thixcin E	5
Camel-Carb®	350

	No. 3	No. 4
	(Polyester/Resin)	
Polylite 8130	522	522
Thixcin E	20	5
Cobalt Naphthenate (6%)	10	10
Drikalite AC	925	925
Lupersol Delta	8	8

Caution:

The **Lupersol Delta** and the cobalt naphthenate should never be mixed directly together since a violent reaction may result.

Tile Grout

Formula No. 1

(Latex Base/Acrylic Resin)

1. Oleic Acid	15.0
2. Triethanolamine	8.0
3. **Rhoplex® AC-61** (46.5% N.V.)	365.0
4. **Nopco® NXZ**	3.2
5. **Triton® X405**	9.0
6. **Troysan PMA-30**	3.0
7. **Silane**	0.6
8. Rutile Titanium Dioxide	10.0
9. **Duramite**	1160.0
10. **Carbopol® 934**	2.0
11. Ethylene Glycol	22.0

Procedure:

Premix (1) and (2), stirring until gel (soap) formation is complete; hold aside for later use. Charge main mixing vessel with all **Rhoplex® AC-61** and **Nopco® NXZ** (defoamer). Start mixer, and carefully add premixed (1) and (2), **Triton® X-405** (secondary dispersant and binder stabilizer), **PMA-30** (mildew inhibitor) and **Silane** (adhesion promoter). When all these wet ingredients are well-mixed, add pigments (8) and (9). Mix thoroughly until pigments are well-dispersed (30–60 min, depending on equipment and batch size). The previously mixed and swollen **Carbopol® 934** (viscosity gellant) and ethylene glycol is now added with continued mixing. Grout will gradually increase in viscosity. Continue mixing until compound is homogeneous and then package immediately.

Note:

If antifreeze feature is not indicated, up to $^1/_2$ of the ethylene glycol in final step can be replaced with water.

No. 2

(Cellulose Gum)

White Portland Cement (4 bags @ 94 lb)	376
Drikalite	600
Calcium Chloride (anhyd. powder)	16
Methocel® 65 Hg (400 cps)	8

Note:

For best results in making the dry tile grout the **Methocel®** should be sifted *very slowly* into the other ingredients while being very rapidly agitated.

Foamed Polystyrene Adhesive

Formula No. 1

(Rubber/Resin)

Kraton® D 1116	100
Sylvatac® 105	180
Mistron Vapor®	340
Butyl Zimate®	3
Plastanox® 2246	2
Epon® 1002	3
Cab-O-Sil®	30
Hexane	280
Cyclohexane	70

No. 2

(Rubber/Resin)

Kraton® D-1116	100
Pentalyn® H	120
Cumar® R16	60
Butyl Zimate®	3
Plastanox® 2246	2
Epon® 1002 F	3
Mistron Vapor®	332
Cab-O-Sil®	30
n-Hexane	280
Cyclohexane	70

Paper-Laminated Polystyrene Adhesive

(Polyvinyl Propionate)

Propiofan 5 D	100
Chalk (finely divided)	75
Latekoll® D (4% sol'n.)	5

Note:

Such materials are supplied in rolls and have pronounced internal stresses; therefore, the adhesive must have very good grab.

The adhesive should be applied to the wall with a serrated trowel.

Adhesives for Elastomers and Plastics

Formula No. 1

(Polychloroprene)

Elastomer	100
Light Calcined Magnesia	4
Zinc Oxide	5
Chlorinated Rubber Resin	15
Coumarone-Indene Resin	15
Toluene	200
Ethyl Acetate	200

No. 2

(Polyurethane/Resin)

Desmocoll 400	100
Chlorinated Rubber Resin	25
Methyl Ethyl Ketone	300

Note:

Immediately prior to use, a quantity of **Vulcabond TX** is added to the adhesive and the solution stirred thoroughly until homogeneous.

The proportion of **Vulcabond TX** to be used depends to a great extent on the degree of adhesion required, the pot life that can be tolerated, and other factors relating to the ambient conditions in which the systems are to be used. Generally, however, dosages in the range of 5–15 parts of **Vulcabond TX** per hundred of rubber are found to be suitable.

The adhesive solution containing the isocyanate is applied to the cleaned, and where appropriate, abraded surfaces and the solvent allowed to evaporate completely. On porous surfaces a second coating of the adhesive may be applied and again allowed to dry completely. The adhesive film is finally activated under an infrared heater for 25–40 s before bringing the treated materials into contact with each other. The bonded substrates are generally allowed to mature for 58 h at room temperature before subjecting them to any strain.

Laminating Plastics Adhesive

(Acrylic)

Rhoplex® LE-1126	100.0
Defoamer (50%)	0.4
Hydroxyethyl Cellulose } Premix	0.7
Cold Water	4.9
Sodium Sesquicarbonate (15%)	6.7

Properties:
Visc. (Brookfield LVF, #4 spindle, 6 rpm)   70,000 cps

Flocking-to-Plastics Adhesive

(Acrylic)

Rhoplex® LE-1126	100.0
Defoamer (50%)	0.4
Hydroxyethyl Cellulose } Premix	0.5
Cold Water	3.5
Sodium Sesquicarbonate (15%)	6.7

Properties:
Visc. (Brookfield LVF, #4 spindle, 6 rpm)   40,000 cps

Plastic Coating and Sealant

(Acrylic)

Rhoplex® LE-1126	100.0
Sodium Bicarbonate	2.5
Calcium Carbonate Slurry (70%)	60.0
Hydroxyethyl Cellulose ⎫ Premix	1.3
Cold Water ⎭	5.0

Properties:
Visc. (Brookfield LVF, #4 spindle, 6 rpm)
50,000 cps

PVC and Nonwovens Sealants

	No. 1	No. 2
	— *(Latex/Resin)* —	
Geon® 460X1	100.00	—
Geon® 460X2	—	100.00
Carbopol® 934	1.25	0.75
M-Pyrol	5.00	5.00
Nalco 123	0.25	0.25
Ammonium Hydroxide	yes	yes

Properties:
Visc. (Brookfield RVF,		
#6 spindle, 20 rpm), cP	24,500	22,000
pH	5.5	7.7

Note:
Compounds applied by engraved (lined) reverse roll coater.

Pigmented, Plasticized PVC Sealant

(Vinyl Acetate Ethylene)

Airflex 400	100
Plasticizer*	0–20

---

* Dibutyl phthalate, **Benzoflex 9-88** or **KP 140**

Rubber, PVC, and Metals Adhesive

	Formula No. 1	No. 2
	(Rubber)	
Natural Rubber (RSS 1)	50.0	50.0
SBR 1712	50.0	50.0
Cofill 11	6.0	—
Hexamethylene Tetramine	1.6	—
Ultrasil VN 3	12.0	—
Durex O	28.0	45.0
Zinc Oxide RS	5.0	5.0
Stearic Acid	3.0	1.5
Rosin	1.5	1.5
Phenyl-beta-Naphthylamine	1.0	1.0
Softener (high aromatic)	12.0	12.0
CBS	1.1	1.1
MBTS	0.3	0.3
Sulfur	3.0	3.0

Vulcanization: 40 min at 145 C.

Metal Cement

(Water-Based)

1. Potassium Hydroxide	4.5–5
2. Silica Hydrogel	15.2–18.8
3. Methyl Trimethoxysilane	1.12–1.23
4. Water	to make 100.0

Procedure:

Mix #1–#3 and heat under pressure then add #4 with mixing.

Metal Coating

(Rubber)

Kraton® 1101	100.0
Antioxidant 330	0.3
DLTDP	0.3

Properties:

180° Peel Strength on Mild Steel	1.1 pli
180° Peel Strength on Glass	0.9 pli
No surface tack	

Note:

Adhesion may be increased by adding a coumarone-indene resin (30 parts doubles adhesion). Strippability may be increased by adding a release agent (0.5 parts **Armid O** halves adhesion). **Kraton® 1102** may be substituted for lower solution viscosity at the expense of high-temperature performance. Soft clay, talc, whiting, or carbon black may be used to extend the compound.

Uses:

Protective coating—tank lining, finished metal parts, mild steel. The formulation has been used as a protective lining for a tank containing 10% HCl for 3 yrs without failure.

Fiberglass, Metals, and Plastics Adhesive

(EVA/PVC)

Geon® 577 (56%)	89
Picco® A-60 (55%)	91
Calcium Carbonate (70%)	43
Antimony Oxide (70% disp.)	7
Good-Rite® K-718 (12.5%)	16

Metal-to-Metal Adhesive

	Formula No. 1	No. 2	No. 3
	(Acrylonitrile/Butadiene)		
Chemigum N-5	100.0	100.0	100.0
SP-8010	70.0	—	—
SP-8014	—	70.0	—
SP-8214	—	—	100.0
Methyl Ethyl Ketone	510.0	510.0	510.0
Sulfur	2.0	2.0	2.0
Zinc Oxide	5.0	5.0	5.0
Altax	1.0	1.0	1.0

Cure 20–40 min at 320 F.

Metal Adhesives

Formula No. 1

(Acrylate)

C-3000	60
SR-238	23
SR-285	14
Triacrylate	5
FC-430	1
Irgacure 184 or **651**	2

No. 2

(Acrylate)

C-3000	55
SR-238	23
SR-285	14
Triacrylate	10
FC-430	1
Irgacure 184 or **651**	2

No. 3

(Acrylate)

C-9502	65
SR-238	26
SR-285	6
Triacrylate	5
FC-430	1
Irgacure 184 or **651**	2

No. 4

(Acrylate)

C-9502	60
SR-238	26
SR-285	6
Triacrylate	10
FC-430	1
Irgacure 184 or **651**	2

No. 5

(Acrylate)

C-3000	50
C-2000	25
Triacrylate	10
SR-285	12
FC-430	1
Irgacure 651	2

No. 6

(Acrylate)

C-3000	50
SR-344	25
Triacrylate	10
SR-285	12
FC-430	1
Irgacure 651	2

No. 7

(Acrylate)

C-3000	50
SR-285	14
Diacrylate	33
FC-430	1
Irgacure 651	2

Tube-Winding Adhesive

Formula No. 1

(Water-Based, Dextrin)

Water	48.5
Defoamer	0.3
Preservative	0.1
Stadex 128	46.0
Sodium Hydroxide (50% sol'n.)	0.9
Borax (5 mole)	4.2

Procedure:
Add the first five ingredients in the order listed with good agitation.
Mix well. Heat to 160 F. Add the borax. Heat to 170 F. Hold 30 min
with agitation. Turn off agitator and let foam come to top. Draw off hot
adhesive from bottom of tank to avoid incorporating additional foam.

Properties:

Solids (Refractometer)		50–52%
pH		≈ 9
Visc. (Brookfield)	@ 150 F	≈ 1,500 cps
	@ 80 F	≈ 30,000 cps

No. 2

(Water-Based, Dextrin)

Water	52.0
Clay (inert **ASP®-400** or equivalent)	8.0
Stadex 45B	40.0

Procedure:
Mix at room temperature. Cook 20 min at 195 F. Cool below 140 F.
Apply at 110–120 F.

Properties:

Solids (Refractometer)	41–42%
pH	≈ 9.2
Visc. (Brookfield) @ 110 F	≈ 1000 cps

No. 3

(Water-Based, Dextrin)

Water	40.8
Preservative	0.1
Defoamer	0.1
Stadex 128	21.0
Stadex 140	10.0
Sodium Silicate (38% solids sol'n.)	26.0

Procedure:
Mix the first five ingredients at room temperature. Heat to 195 F and hold for 15 min. Cool below 130 F. Adjust the pH to 10.9–11.2 with a 50% sodium hydroxide sol'n. (approximately 1% of adhesive weight). Carefully mix in the sodium silicate sol'n.

Properties:
Solids (Refractometer)	40–41%
pH	11.0–11.2
Visc. (Brookfield) @ 80 F	≈ 700 cps

No. 4

(Water-Based, Dextrin)

Water	53
Stadex 53B	47

Procedure:
Mix at room temperature. Heat to 200 F. Hold @ 200 F for 30 min. Cool. Dilute as required.

Properties:
Solids (Refractometer)	44–45%
Visc. (Brookfield) @ 80 F	≈ 2000 cps

Metal-to-Glass Adhesive

(Solvent-Based)

Collodion Cotton (wetted with 50% alcohol)	19.6
Ethyl Acetanilide	3.5
Tricresyl Phosphate	1.7
Asbestine Powder	28.0
Titanium Dioxide	1.0
Solvents (mixture consisting of easily volatile esters and methanol)	46.2

Note:
When small parts are bonded, the adhesive is spread on the adherent, is allowed to be absorbed for a few seconds until the adhesive layer has become highly viscous, and then the piece is pressed on.

Paper-to-Aluminum Laminating Adhesive

Formula	No. 1	No. 2	No. 3	No. 4	No. 5
			(Acrylic/Resin)		
Hycar® 2600X83	100.0	100.0	50.0	75.0	75.0
Carboxylated Polyvinyl Acetate	—	—	50.0	25.0	25.0
Borated Casein	10.0	—	—	—	—
Dibutyl Phthalate	—	—	10.0	—	—
Carbopol® 934	1.5	1.5	1.5	1.5	2.0
Ammonium Hydroxide			to pH 9.0		

Procedure:

Blend the compounding ingredients and raise the pH of the blend to 9.0 with 10% ammonium hydroxide. Add the **Carbopol®** resin and stir at medium speed until the blend thickens. Increase mixing speed and readjust to a final compound pH of 9.0.

Aluminum Foil Adhesive

Formula No. 1

(Vinyl Acetate/Ethylene)

Airflex 400	75
Airflex 500	25
Dibutyl Phthalate	5–10

No. 2

(Vinyl Acetate/Ethylene)

Airflex 400	100
Dibutyl Phthalate	5–10

Note:

If the adhesive does not wet the film because of exuded slip agent or plasticizer, etc., the addition of a wetting agent may help.

Hot-Melt Adhesive for Aluminum Foil No. 9

(Polyethylene/Resin)

Polyethylene (low m.w.)	37
Ethylene/Ethyl Acrylate Copolymer	20
Eastman H-130	25
Polywax 1000	11

Hot-Melt Adhesive for Aluminum Foil No. 10

(Polyethylene/Resin)

Polyethylene (low m.w.)	64
Eastman H-130	25
Polywax 1000	11

Hot-Melt Adhesive for Bonding Polyethylene (or Polypropylene) to Aluminum

(Vinyl Acetate/Resin)

Elvax 4310	28.5
Elvax 4260	19.0
Nirez® V-2040	38.0
Victory® White	14.2
Antioxidant	0.3

Properties:

Visc. (Brookfield Thermal)	@ 325 F	26,000 cps
	@ 350 F	13,000 cps
Softening Pt. (Ring & Ball)		95 C

Chapter III

HOT-MELT ADHESIVES

Hot-Melt, Pressure-Sensitive Tape Adhesive

Formula No. 1

(Rubber/Resin)

Kraton® D1107	100
Escorez® 1310	140
Shellflex® 371	10
Antioxidant	2

	No. 2	No. 3	No. 4	No. 5
	(Styrene-Butadiene/Resin)			
Solprene 418	100	100	100	100
Wingtack® 95	150	150	150	150
Amoco® Resin 18-210	40	—	—	—
Amoco® Resin 18-290	—	40	40	—
Amoco® Polybutene L-14	10	—	80	—
Amoco® Polybutene H-300	—	—	—	80
Amoco® Polybutene H-1500	—	80	—	—
Irganox® 1010	2	2	2	2
Properties:				
Rolling Ball Tack, PSTC-6, cm	—	—	1.3	—
Quick Stick Adhesion, PSTC-5, lb	11.6	4.3	—	4.1
Peel Adhesion, PSTC-1, lb	10.7	13.7	—	11.3
Shear Adhesion, Amoco Test, s	42.8	28.9	—	135.0

Hot-Melt, Pressure-Sensitive
High-Hold Tape Adhesive

(Rubber/Resin)

Kraton® GX 1657	100.00
Regalrez® 1076	122.00
Irganox® 1010	1.00
Tinuvin® 327	0.25
Tinuvin® 770	0.25

Properties:

Rolling Ball Tack	5 cm
Polyken® Probe Tack	0.7 kg
Loop Tack	3.5 pli
Loop Tack	610 N/m
Finger Tack	Good
180° Peel	3.5 pli
180° Peel	610 N/m
Holding Power to Steel*	4000 min
Holding Power to Kraft*	2000 min
SAFT to Mylar	90 C
SAFT to Kraft	60 C
Melt Visc. @ 177 C	80 Pa·s

---

* Time for disbonding of 13 × 13 mm bond under 2-kg load at 25 C.

Hot-Melt, Pressure-Sensitive Office Tape Adhesive

Formula No. 1

(Rubber/Resin)

Kraton® GX 1657	100.00
Regalrez® 1076	57.00
Regalrez® 1018	129.00
Irganox® 1010	1.00
Tinuvin® 327	0.25
Tinuvin® 770	0.25

Properties:
Rolling Ball Tack	2.5 cm
Polyken® Probe Tack	0.4 kg
Loop Tack	3.5 pli
Loop Tack	610 N/m
Finger Tack	Very good
180° Peel	2.7 pli
180° Peel	470 N/m
Holding Power to Steel*	50 min
Holding Power to Kraft*	10 min
SAFT to Mylar	60 C
SAFT to Kraft	45 C
Melt Visc. @ 177 C	25 Pa·s

---

*   Time for disbonding of 13 × 13 mm bond under 2-kg load at 25 C.

No. 2

Kraton® **GX 1657**	100.00
Arkon® **P90**	135.00
Indopol® **H100**	68.00
Irganox® **1010**	1.00
Tinuvin® **327**	0.25
Tinuvin® **770**	0.25

Properties:
Polyken® Probe Tack	1.1 kg
Loop Tack	3.0 pli
Loop tack	530 N/m
180° Peel	2.5 pli
180° Peel	440 N/m
Holding Power	1 h
Service Temp.*	65 C
Melt visc. @ 177 C	25 Pa·s

---

*   Temperature at which a 25 × 25 mm lap shear bond to **Mylar**® fails under a 1-kg load in a cabinet whose temperature is raised at a rate of 22 C/h.

Hot-Melt, Pressure-Sensitive
High-Tack Permanent Tape Adhesive

(Rubber/Resin)

Kraton® GX 1657	100.00
Regalrez® 1076	110.00
Regalrez® 1018	76.00
Irganox® 1010	1.00
Tinuvin® 327	0.25
Tinuvin® 770	0.25

Properties:

Rolling Ball Tack	2 cm
Polyken® Probe Tack	1.2 kg
Loop Tack	6 pli
Loop Tack	1050 N/m
Finger Tack	Very good
180° Peel	4.0 pli
180° Peel	700 N/m
Holding Power to Steel*	200 min
Holding Power to Kraft*	80 min
SAFT to Mylar	70 C
SAFT to Kraft	55 C
Melt Visc. @ 177 C	30 Pa·s

---

*   Time for disbonding of 13 × 13 mm bond under 2-kg load at 25 C.

Hot-Melt, Pressure-Sensitive
Low-Peel Repositionable-Tape Adhesive

(Rubber/Resin)

Kraton® GX 1657	100.00
Regalrez® 1076	54.00
Regalrez® 1018	68.00
Irganox® 1010	1.00
Tinuvin® 327	0.25
Tinuvin® 770	0.25

Properties:

Rolling Ball Tack	2 cm
Polyken® Probe Tack	0.5 kg
Loop Tack	2.5 pli
Loop Tack	440 N/m
Finger Tack	Very good
180° Peel	2.5 pli
180° Peel	440 N/m
Holding Power to Steel*	300 min
Holding Power to Kraft*	20 min
SAFT to Mylar	75 C
SAFT to Kraft	50 C
Melt Visc. @ 177 C	60 Pa·s

*   Time for disbonding of 13 × 13 mm bond under 2-kg load at 25 C.

Cured Adhesive for Heat-Resistant Masking Tape

(Rubber/Resin)

Zonarez B-115	100.0
Smoked Sheet Rubber	75.0
SBR-24% Styrene	25.0
Zinc Oxide	50.0
Alkyl Phenol Sulfide Curing Agent*	8.0
Mineral Oil	3.0
Antioxidant	1.0
Heptane	350.0

*   Cure at 300–350 F

Hot-Melt Bonding-Tape Adhesive

(Isobutylene/Isoprene)

Chlorobutyl HT 10-66	100.0
SRF Black	80.0
Amberol ST 140F	7.0
Pentalyn® K	3.0
Flexon 745	10.0

Polyethylene AC-617	5.0
Stearic Acid	4.0
Zinc Oxide	10.0
Permalux	3.0

Cure: 15–30 s @ 375 F.

Hot-Melt, Pressure-Sensitive, Double-Faced Tape Adhesive

(Vinyl Isobutyl Ether)

Lutonal IC	9
Lutonal I 30	30
Lutonal I 60	120
Carbigen K90	5

Note:

This is excellent for double-faced tapes with high coating weight on a textile carrier, e.g., for securing edges (carpet-laying tape).

Hot-Melt, Pressure-Sensitive, Medical-Tape Adhesive

(Vinyl Isobutyl Ether)

Lutonal IC	10.0
Lutonal I 60	60.0
Lutonal I 30	5.0
Antioxidant 2KF	0.5
Zinc Oxide	25.0
Mineral Spirits (b.p. 75–95 C)	≈ 80.0

Note:

Pressure-sensitive tapes and films produced with hot-melt adhesive compounds consisting of 40–60 parts **Lutonal I 60** and 40–60 parts bitumen can be used as "bituminous adhesive off the roll" for a great variety of applications. They can, for instance, be used on building sites where cookers for hot bituminous adhesives are undesirable because of the danger of fire and accidents, or for other reasons.

Hot-Melt Plasticized Decorative PVC Film

(Vinyl Isobutyl Ether/Resin)

Acronal 80 D	80
Lutonal I 65	20

Hot-Melt Protective Coating

(Rubber/Resin)

Kraton® 1102	100.0
Foral® 105	100.0
Shellwax 300	85.0
Armid O	0.5
Antioxidant 330	0.5
DLTDP	0.5

Properties:
Melt visc. @ 325 F   3600 cps

Note:
Adhesion may be improved by removing the release agent, **Armid O.**

Hot-Melt High-Gloss Coating

(Ethylene Vinyl Acetate)

Ultrathene UE 612-04 (19% VA, 150 MI)	20
Polyterpene Resin (115 C)	15
Paraffin Wax (m.p. 150 F)	60
Microcrystalline Wax (m.p. 155 F)	5

Note:
This is useful for difficult applications involving plastic films and metal foils.

Hot-Tack Coating for Nonporous Surfaces

(Ethylene Vinyl Acetate/Resin)

Zonarez 7115	25.0
Elvax 4355	35.0
Paraffin Wax (m.p. 155 F)	35.0
Microcrystalline Wax (m.p. 180 F)	5.0

Properties:

Visc. @ 300 F	11,100 cps
@ 350 F	5,130 cps
@ 390 F	1,513 cps
Adhesion:	
Kraft-to-Kraft	38 oz/in.
Kraft-to-PE	10 oz/in.
Mylar	12 oz/in.

Note:

This formulation has excellent hot tack under a broad range of conditions making it suitable for use on high-speed form-fill machines. It also has shown excellent adhesion to substrates such as polypropylene, polyester, and aluminum foil without sacrificing those physical properties necessary for flexible packaging.

Hot-Melt Coatings and Adhesives

	Formula No. 1	No. 2	No. 3	No. 4
	(Ethylene Vinyl Acetate/Resin)			
EVA (28% VA)*	33.3	33.3	33.3	33.3
Paraffin Wax (≈ m.p. 145 F)	33.4	33.4	33.4	33.4
Nevtac 80	33.3	—	—	—
Nevtac 100	—	33.3	—	—
Nevtac 115	—	—	33.3	—
Super Nevtac 99	—	—	—	33.3

---

* EVA 507, Elvax 260 or UE 634-04

Formula No. 1	No. 2	No. 3	No. 4
	(cont'd.)		

Properties:

Cloud Point, C	68	71	200	56
Visc. @ 121 C, cps	11,700	12,400	17,200	12,750
@ 177 C, cps	4,560	4,780	5,560	5,000
Tensile Strength, psi	540	641	756	589
Elongation, %	91	74	64	82

	No. 5	No. 6	No. 7	No. 8
	(Ethylene Vinyl Acetate/Resin)			
EVA (18% VA)*	33.3	33.3	33.3	33.3
Wax	33.4	33.4	33.4	33.4
Nevtac 80	33.3	—	—	—
Nevtac 100	—	33.3	—	—
Nevtac 115	—	—	33.3	—
Super Nevtac 99	—	—	—	33.3

Properties:

Cloud Point, C	57	57	62.	57
Visc. @ 121 C, cps	1020	1190	1240	1140
@ 177 C, cps	258	292	298	285
Tensile Strength, psi	490	589	670	551
Elongation, %	35	38	28	38

---

* **Elvax 410 or UE 649-04**

	No. 9	No. 10
	(Ethylene Vinyl Acetate/Resin)	
EVA (27–28% VA, 22–28 MI)	33.3	33.3
Super Nirez™ 5100	33.3	—
Super Nirez 5120	—	33.3
Paraffin Wax (m.p. 150 F)	33.3	33.3

Properties:
  Color (Gardner)

Initial	< 1.0	< 1.0
96 h @ 176.6 C	12.5	12.5
Cloud Pt., C	60	60

	No. 11	No. 12
	(Ethylene Vinyl Acetate/Resin)	
EVA (27–28% VA, 134–138 MI)	50	50
Super Nirez™ 5100	50	—
Super Nirez™ 5120	—	50
Properties:		
Cloud Pt., C	32	32

No. 13

(Ethylene Vinyl Acetate/Resin)

EVA (MI 2–3)	25.0
Micro Wax (m.p. 180 F)	25.0
Epolene C-15	12.5
WingTack® 115	37.5
WingStay L	0.5

No. 14

(Ethylene Vinyl Acetate/Resin)

Piccotex® 120	15
Elvax 250	15
Paraffin (m.p. 150 F)	45
Micro Wax (m.p. 180 F)	25

No. 15

(Ethylene Vinyl Acetate/Resin)

Piccotex® 120	15.0
Elvax 260	15.0
Paraffin (≈ m.p. 155 F)	45.0
Micro Wax (≈ m.p. 180 F)	25.0
Tenox BHY	0.1
Armoslip E	0.1

	No. 16	No. 17	No. 18
	(Ethylene Vinyl Acetate/Resin)		
Piccotex 120	15	15	—
EVA (28% VAc; 150 MI)	15	15	—
Wax (≈ m.p. 150 F)	70	70	—
Neville R-12A	—	—	50
Shellflex®	—	—	25
Antioxidant 330	—	—	1
Heat to 350 F. Add:			
Kraton® 1101	—	—	25
Stir until dissolved.			

	No. 19	No. 20	No. 21	No. 22	No. 23	No. 24
	——— *(Ethylene Vinyl Acetate/Resin)* ———					
ST-5115	15.0	15.0	15.0	17.0	18.0	20.0
Elvax 220	15.0	—	—	—	—	—
Elvax 250	—	15.0	—	—	—	—
Elvax 260	—	—	15.0	25.0	30.0	30.0
Paraffin Wax						
(≈ m.p. 150 F)	70.0	45.0	70.0	58.0	52.0	50.0
Microwax						
(≈ m.p. 180 F)	—	25.0	—	—	—	—

	No. 25	No. 26	No. 27	No. 28	No. 29
	—— (Ethylene Vinyl Acetate/Resin) ——				
Paraffin Wax (≈ m.p. 154 F)	70	66.5	63.0	59.5	56.0
Elvax 250	30	28.5	27.0	25.5	24.0
Nevex 100	—	5.0	10.0	15.0	20.0

	No. 30	No. 31	No. 32	No. 33
	(Ethylene Vinyl Acetate/Resin)			
Paraffin Wax (≈ m.p. 151 F)	50	33	50	50
EVA (28% VAc; med. MI)	17	33	30	—
EVA (28% VAc; low MI)	—	—	—	30
Klyrvel 90	33	33	20	20

	No. 34	No. 35	No. 36
	(Pressure-Sensitive, Rubber/Resin)		
Kraton® 1107	100.0	100.0	100.0
Wingtack® 95	100.0	100.0	100.0
Irganox® 1010	1.5	1.5	1.5
Amoco® Polybutene	10–80	10–80	10–80
Amoco® Resin 18-210	—	20–80	—
Amoco® Resin 18-290	—	—	20–80
Cumar® LX-509	20–80	—	—

	No. 37
	(Rubber/Resin)
Kraton® 1107	66.6
Kraton® 1102	33.3
Super Sta-Tac® 80	60.0
Sta-Tac® 100	20.0
Betaprene® AC-130	40.0
Antioxidant (nonstaining)	1.5

Properties:
Visc. (Brookfield Thermosel)

@ 300 F	50,000 cps
@ 350 F	32,000 cps

1.0 mil coated on polyester film:

Rolling Ball Tack	4–5 in.
180° Peel Adhesion	22–2400 g
Creep Resistance @ room temp. at 2 psi	100 h—no failure
Creep Resistance @ 140 F under 1 psi load	2 h —no failure

No. 38

(Rubber/Resin)

Kraton® 1107	100.0
Super Sta-Tac® 80	70.0
Antioxidant (nonstaining)	1.0

Properties:
Visc. (Brookfield Thermsel)

@ 300 F	32–34,000 cps
@ 350 F	25–26,000 cps
Rolling Ball Tack	0.7 in.
Peel Adhesion (to stainless steel)	1800–2000 g

1.0 mil on polyester film:

Creep Resistance @ 140 F under 1 psi load	2 h—no failure
Creep Resistance @ room temp. at 1 psi	148 h—no failure
Quick Stick (loop)	1500 g

	No. 39	No. 40	No. 41
		(Rubber/Resin)	
Kraton® D 1107	100	—	—
Kraton® D 1111	—	100	100
Escorez® 1310	140	140	140
Shellflex® 371	10	10	10
Piccovar® AP-33	—	—	40
Antioxidant	2	2	2

Properties:

Rolling Ball Tack, cm	1.5	2.8	2.2
Polyken® Probe Tack, kg	1.75	1.65	1.92
SAFT* to Mylar, C	93	105	88
SAFT to Kraft, C	78	78	74
Holding Power**			
Kraft, min	2000	7000+	3100
Steel, min	660	7000+	3000
180° Peel,			
pli	6.2	6.4	7.1
N/m	1090	1120	1240
Toluene Sol'n. Visc. @ 40% solids, Pa·s	0.59	0.47	0.28

---

* Temperature at which a 25 mm × 25 mm lap shear bond fails under a 1-kg load in a cabinet whose temperature is raised at a rate of 22 C/h.

** PSTC Method No. 7, 13 mm × 13 mm contact area to test substrate with a 2-kg load.

	No. 42	No. 43	No. 44	No. 45	No. 46
			(Rubber/Resin)		
Kraton® D 1107	100	—	—	100	—
Kraton® D 111	—	100	100	—	100
Wingtack® 95	150	150	150	100	100
Cumar® LX-509	—	—	—	40	40
Tufflo® 6056	10	10	60	60	60
Antioxidant	2	2	2	2	2

Properties:

	No. 42	No. 43	No. 44	No. 45	No. 46
Melt visc., Pa·s	116	>2000	1430	55	>2000

	No. 47	No. 48	No. 49	No. 50	No. 51
			(Rubber/Resin)		
Kraton® 1107	100	100	100	100	100
Wingtack® 95	100	100	100	100	100
Amoco® Resin					
18-210	—	—	—	—	50

	No. 47	No. 48	No. 49 *(cont'd.)*	No. 50	No. 51
Amoco® Resin 18-290	40	40	20	40	—
Amoco® Polybutene L-14	60	—	—	—	—
Amoco® Polybutene H-100	—	10	—	—	—
Amoco® Polybutene H-1500	—	—	40	40	20
Irganox® 1010	1	1	1	1	—
Antioxidant 330	—	—	—	—	5
Properties:					
Rolling Ball Tack, TSTC-6, cm	2.3	—	—	—	19.3
Quick Stick Adhesion, PSTC-5, lb	2.0	1.8	5.3	4.5	—
Peel Adhesion, PSTC-1, lb	2.5	5.8	10.7	12.0	2.4
Shear Adhesion, Amoco Test, s	1.6	263.0	37.6	32.3	3309

	No. 52	No. 53 *(Rubber/Resin)*	No. 54
Kraton® 1107	100.0	100.0	100.0
Nevtac 80	100.0	—	—
Nevtac 100	—	100.0	—
Super Tevtac 99	—	—	100.0
Butyl Zimate®	2.0	2.0	2.0
Shellflex® 371		—— 0 or 25.0 ——	
Properties: (no oil)			
Rolling Ball Tack, in.	0.4	3.9	0.8
Probe Tack, kg	1.09	1.40	1.33
180° Peel, oz/in.	70	70	70
Shear Adhesion to Kraft, min	800	2700	3000
Shear Adhesion Failure Temp. (Kraft), F	185	184	179
Visc. @ 350 F, cps	138,000	200,000	155,000

Properties: (25 phr oil)

Rolling Ball Tack, in.	0.3	0.6	0.4
Probe Tack, kg	0.77	1.17	0.95
180° Peel, oz/in.	51	62	50
Shear Adhesion to Kraft, min	25	120	50
Shear Adhesion Failure Temp. (Kraft), F	166	168	167
Visc. @ 350 F, cps	51,500	63,800	53,700

	No. 55	No. 56
	(Rubber/Resin)	
Kraton® GX 1657	100.00	100.00
Regalrez® 1078	122.00	110.00
Regalrez® 1018	—	76.00
Irganox® 1010	1.00	1.00
Tinuvin® 327	0.25	0.25
Tinuvin® 770	0.25	0.25

Properties:

Polyken® Probe Tack, kg	0.7	1.2	1.1
Loop Tack, pli	3.5	4.0	3.0
Loop tack, N/m	610	700	530
180° Peel, pli	3.0	3.5	2.5
180° Peel, N/m	530	610	440
Holding Power, h	> 50	3	1
Service Temp.*, C	90	70	65
Melt visc. @ 177 C, Pa·s	100	30	25

---

* Temperature at which a 25 × 25 mm lap shear bond to **Mylar®** fails under a 1-kg load in a cabinet whose temperature is raised at a rate of 22 C/h.

	No. 57	No. 58	No. 59	No. 60
		(Rubber/Resin)		
American Gilsonite Selects	40	40	35	35
Paraffin (m.p. 135 F)	35	25	25	25
Emerez® 1532	10	10	10	10
Dibutyl Phthalate	15	15	—	—
Carlisle Wax 240	—	10	—	10
Wood Rosin	—	—	20	—
Mineral Oil	—	—	10	10

	No. 61	No. 62	No. 63	No. 64
		(Rubber/Resin)		
American Gilsonite Selects	50	40	40	55
Paraffin (m.p. 135 F)	35	30	30	30
Mineral Oil	15	10	10	—
Piccopale® 100 or 70	—	20	—	—
Piccolastic A-25	—	—	20	—
Hercolyn® D	—	—	—	15

	No. 65	No. 66	No. 67	No. 68
		(Rubber/Resin)		
American Gilsonite Selects	35	40	40	40
Paraffin (m.p. 135 F)	30	30	30	30
Mineral Oil	15	10	10	10
Pentalyn® G	20	—	—	—
Wingtack® 95	—	20	—	—
Piccotex® 120	—	—	20	—
Stikvel P	—	—	—	20

	No. 69	No. 70	No. 71	No. 72
		(Rubber/Resin)		
American Gilsonite Selects	40	40	40	40
Paraffin (m.p. 135 F)	30	30	30	30
Mineral Oil	10	10	10	10
Stikvel W	20	—	—	—
Staybelite® Ester 10	—	20	—	—
Sta-Tac® B	—	—	20	—
Sta-Tac® 100	—	—	—	20

	No. 73	No. 74	No. 75	No. 76
		(Rubber/Resin)		
American Gilsonite Selects	40	35	35	35
Paraffin (m.p. 135 F)	30	25	25	25
Mineral Oil	10	10	10	10
Wood Rosin	20	—	—	—
Epolene C-15	—	10	10	10
Carlisle Wax 240	—	20	—	—
Dibutyl Phthalate	—	—	20	—
Santocizer 160	—	—	—	20

	No. 77	No. 78	No. 79	No. 80	No. 81
		(Rubber/Resin)			
American Gilsonite Selects	35	30	30	40	35
Paraffin (m.p. 135 F)	20	25	25	30	25
Mineral Oil	15	15	15	—	—
Emerez 1532	—	10	10	10	10
Dibutyl Phthalate	—	—	—	20	—
Santocizer 160	—	—	20	—	—
Castor Wax	20	—	—	—	—
Pentalyn® G	—	20	—	—	—
Carlisle Wax 240	—	—	20	—	—

No. 82

(Rubber/Resin)

Kraton® 1107	100.0
Wingtack® 95	100.0
Irganox® 1010	1.5
Polybutene	10–80
Amoco® Resin 18-210	20–80

No. 83

(Latex/Resin)

	Dry	Wet
Dow Latex XD-30223 (42%)	70.00	166.67
Coumarone-Indene Resin Emulsion (50%)	10.00	20.00
Hydrocarbon Resin Emulsion (62.5%)	10.00	16.00
Butyl Benzyl Phthalate (100%)	10.00	10.00
Polyacrylate Thickener (11%)	0.45	4.00

	*No. 84	No. 85	No. 86
		(Rubber/Resin)	
Kraton® GX-1657	100	100	100
Arkon® P85	123	104	91
Indopol® H100	27	46	22
Properties:			
Polymer Content, %w	40	40	47
Midblock Phase TG, C	–10	–19	–16
Rolling Ball Tack, cm	7	2.1	3
Polyken® Probe Tack, g	720	840	390
Loop Tack, pli	6.1	4.9	4.7
Loop Tack, N/m	1090	860	820
Finger Tack	Fair+	Good+	Good–

---

* Formulations also contain 1 phr **Irganox®** 1010, 0.25 phr **Tinuvin®** 327, and 0.25 phr **Tinuvin®** 770.

180° Peel, pli	4.1	3.4	3.2
180° Peel, N/m	720	600	560
Holding Power to Steel, min	1400	105	1250
Holding Power to Kraft, min	65	4	70
SAFT to Mylar, F	182	177	191
SAFT to Mylar, C	83	81	88
SAFT to Kraft, F	127	107	144
SAFT to Kraft, C	53	42	62
Melt Visc. @ 177 C, Pa·s	47	44	84

	*No. 87	No. 88	No. 89
		(Rubber/Resin)	
Kraton® GX-1657	100	100	100
Arkon® P85	96	106	143
Arkon® P125	7	—	—
Indopol® H100	4	—	7
Properties:			
Polymer Content, %w	50	47	40
Midblock Phase TG, C	−10	−4	−2
Rolling Ball Tack, cm	7	13	24
Polyken® Probe Tack, g	570	560	570
Loop Tack, pli	4.2	4.6	1.3
Loop Tack, N/m	740	810	230
Finger Tack	Fair	Poor	Fair
180° Peel, pli	3.2	3.9	4.9
180° Peel, N/m	560	680	860
Holding Power to Steel, min	> 4000	> 4000	> 4000
Holding Power to Kraft, min	700	> 4000	2100
SAFT to Mylar, F	189	193	190
SAFT to Mylar, C	87	89	88
SAFT to Kraft, F	134	146	136
SAFT to Kraft, C	57	63	58
Melt Visc. @ 177 C, Pa·s	79	104	65

---

* Formulations also contain 1 phr **Irganox®** 1010, 0.25 phr **Tinuvin®** 327, and 0.25 phr **Tinuvin®** 770.

	*No. 90	No. 91	No. 92
		(Rubber/Resin)	
Kraton® GX-1657	100	100	100
Arkon® P85	170	168	135
Indopol® H100	33	64	68
Properties:			
Polymer Content, %wt.	33	30	33
Midblock Phase TG, C	−4	−10	−16
Rolling Ball Tack, cm	21	4	2.0
Polyken® Probe Tack, g	790	1060	1100
Loop Tack, pli	3.2	6.0	5.8
Loop Tack, N/m	560	1050	1020
Finger Tack	Good	Exc.	Exc.
180° Peel, pli	5.6	4.9	3.7
180° Peel, N/m	980	860	650
Holding Power to Steel, min	1500	240	50
Holding Power to Kraft, min	50	10	3
SAFT to Mylar, F	164	161	164
SAFT to Mylar, C	73	72	73
SAFT to Kraft, F	114	107	103
SAFT to Kraft, C	46	42	40
Melt Visc. @ 177 C, Pa·s	22.2	15.6	21.5

* Formulations also contain 1 phr **Irganox®** 1010, 0.25 phr **Tinuvin®** 327, and 0.25 phr **Tinuvin®** 770.

	*No. 93	No. 94	No. 95
		(Rubber/Resin)	
Kraton® GX-1657	100	100	100
Regalrez® 1076	94	64	68
Regalrez® 1018	56	86	45

* Formulations also contain 1 phr **Irganox®** 1010, 0.25 phr **Tinuvin®** 327, and 0.25 phr **Tinuvin®** 770.

Properties:

Polymer Content, %wt.	40	40	47
Midblock Phase TG, C	−10	−19	−16
Rolling Ball Tack, cm	1.7	4	1.5
Polyken® Probe Tack, g	1100	660	840
Loop Tack, pli	5.8	3.4	3.9
Loop Tack, N/m	1020	600	680
Finger Tack	Exc.	Exc.	Exc.
180° Peel, pli	3.5	2.6	3.1
180° Peel, N/m	610	460	540
Holding Power to Steel, min	460	95	2000
Holding Power to Kraft, min	130	9	170
SAFT to Mylar, F	167	153	171
SAFT to Mylar, C	75	67	72
SAFT to Kraft, F	142	122	139
SAFT to Kraft, C	62	50	59
Melt Visc. @ 177 C, Pa·s	23.1	21.2	45

	*No. 96	No. 97	No. 98
		(Rubber/Resin)	
Kraton® GX-1657	100	100	100
Regalrez® 1076	83	104	123
Regalrez® 1018	17	9	27
Properties:			
Polymer Content, %wt.	50	47	40
Midblock Phase TG, C	−10	−4	−2
Rolling Ball Tack, cm	2.0	2.8	4.2
Polyken® Probe Tack, g	870	950	900
Loop Tack, pli	4.6	4.9	3.8
Loop Tack, N/m	810	860	670
Finger Tack	Exc.	Good+	Exc.

---

* Formulations also contain 1 phr Irganox® 1010, 0.25 phr Tinuvin® 327, and 0.25 phr Tinuvin® 770.

180° Peel, pli	2.8	3.5	3.9
180° Peel, N/m	490	610	680
Holding Power to Steel, min	4000	>4000	710
Holding Power to Kraft, min	1000	3100	330
SAFT to Mylar, F	186	194	172
SAFT to Mylar, C	86	90	78
SAFT to Kraft, F	155	163	135
SAFT to Kraft, C	68	73	57
Melt Visc. @ 177 C, Pa·s	63	54	24.5

	*No. 99	No. 100	No. 101
		(Rubber/Resin)	
Kraton® GX-1657	100	100	100
Regalrez® 1076	131	111	79
Regalrez® 1018	72	120	124

Properties:

Polymer Content, %wt.	33	30	33
Midblock Phase TG, C	−4	−10	−16
Rolling Ball Tack, cm	1.9	2.6	1.2
Polyken® Probe Tack, g	1400	1200	880
Loop Tack, pli	6.6	5.4	5.4
Loop Tack, N/m	1160	920	950
Finger Tack	Exc.	Exc.	Exc.
180° Peel, pli	5.0	3.6	3.6
180° Peel, N/m	880	630	630
Holding Power to Steel, min	210	70	50
Holding Power to Kraft, min	80	15	6
SAFT to Mylar, F	146	134	132
SAFT to Mylar, C	63	57	56
SAFT to Kraft, F	117	115	106
SAFT to Kraft, C	47	46	41
Melt Visc. @ 177 C, Pa·s	9.4	6.4	8.5

---

* Formulations also contain 1 phr **Irganox®** 1010, 0.25 phr **Tinuvin®** 327, and 0.25 phr **Tinuvin®** 770.

	No. 102	No. 103	No. 104
	(Styrene-Butadiene/Resin)		
Ameripol SN 600	85	15	100
Ameripol 1006	15	—	—
Ameripol 1009	—	85	—
Dymerex	30	43	43
Pexate 549	13	—	—
Dixie Clay	20	30	30
Cab-O-Sil®	10–20	10	20
Zinc Oxide	10	—	—
Indicated cps	isopentane	pentane	pentane

	No. 105	No. 106	No. 107
	(Styrene-Butadiene/Resin)		
Solprene 414	60	60	60
Solprene 1205C	40	40	40
Foral® 85	100	110	—
Zonester® 85	—	—	110
AgeRite Geltrol	4	4	4
STDP	1	1	1

Properties:			
Rolling Ball Tack,*			
in.	3.7	5.5	3.1
mm	94	139.7	78.8
180° Peel,*			
lb/in.	3.8	4.1	3.5
kN/m	0.70	0.72	0.61
Visc.			
Original, cps	72,000	58,000	62,000
Aged 24 h @ 350 F (177 C), cps	70,000	60,000	72,000

---

* Measured on 1.5 mil films on 1mil **Mylar®**.

	No. 108	No. 109
	(Styrene-Butadiene)	
Solprene 418	100	100
Wingtack® 95	150	150
Irganox® 1010	2	2
Amoco® Polybutene	0–125	0–125
Amoco® Resin 18-210	—	40
Amoco® Resin 18-290	40	—

No. 110

(Styrene-Butadiene/Resin)

Solprene 1205	60
Solprene 303	40
Staybelite® Ester 10	200
Polycizer 162	40
Polygard	2

	No. 111	No. 112	No. 113	No. 114
	(Styrene-Butadiene/Resin)			
Solprene 418	100	100	100	100
Wingtack® 95	150	150	150	150
Amoco® Resin 18-210	40	—	—	40
Amoco® Resin 18-290	—	40	40	—
Amoco® Polybutene L-14	10	—	80	—
Amoco® Polybutene H-300	—	—	—	80
Amoco® Polybutene H-1500	—	80	—	—
Irganox® 1010	2	2	2	2

	No. 111	No. 112	No. 113	No. 114
		(cont'd.)		

Properties:

	No. 111	No. 112	No. 113	No. 114
Rolling Ball Tack, PSTC-6, cm	—	—	1.3	—
Quick Stick Adhesion, PSTC-5, lb	11.6	4.3	—	4.1
Peel Adhesion, PSTC-1, lb	10.7	13.7	—	11.3
Shear Adhesion, Amoco Test, s	42.8	28.9	—	135.0

	No. 115	No. 116
	(Styrene-Isoprene/Resin)	
Styrene-Isoprene (14/86) Block Copolymer	100	100
Super Nirez 5100	100	—
Super Nirez 5120	—	100
Antioxidant	1	1

Properties: (Samples coated 1.0 mil on 2.0 mil polyester film)

	No. 115	No. 116
Visc. @ 176.6 C, cps	78,000	80,000
Polyken® Probe Tack, g	1275	1350
Peel Adhesion, g/*	2300	2200
Quick Stick (loop test), g	2000	1900
Shear Adhesion 2 kg/in.^{2}, days	7	7

---

* g/in. of width

No. 117

(Styrene-Isoprene/Resin)

Zonarez 7115	40.0
Styrene/Isoprene Block Copolymer	60.0
Antioxidant	0.5

Application temp. 350–390 F

No. 118

(Polybutene/Resin)

Kraton® GX-1657	22.2
Super Nirez™ 5100	44.3
Indopol® H-100	33.3
Irganox® 1010	0.2

Properties:
Visc. (Brookfield Thermosel)

RVT, spindle #27 @ 300 F	24,625 cps
@ 300 F	11,200 cps

PSA Properties:
(Sample coated 1.0 dry mil on 2.0 mil Mylar type-"S")

Rolling Ball Tack	6.0+ in.
180° Peel Adhesion to Stainless Steel	3.0 lb/in. of width
140 F Shear Adhesion, 1 psi load	Failure @ 45 min
0° Shear Adhesion @ 77 F, 4.4 psi load	Failure @ 39 min
Quick Stick (loop test)	2.4 psi
Polyken® Probe Tack	
(speed—0.1 cm/s; dwell—instant; weight—"A")	581 g

	No. 119	No. 120
Thermoplastic Elastomer	100	100
Tackifier	100	100
Antioxidant	1	1
Indopol® Polybutene L-14	60	—
Indopol® Polybutene H-1500	—	40
Amoco® Resin 18-290	40	40

Procedure:
All materials except the elastomer are heated to 191 C (375 F) under a nitrogen blanket, and mixed until uniform. The elastomer is added slowly and mixed until homogeneous.

Properties:

		No. 119	No. 120
Visc. @ 177 C (350 F), cps		15,000	42,000
Rolling Ball Tack,	cm	2.3	> 30
	in.	0.9	> 12

Quick Stick Adhesion,	kN/m	0.4	0.8
	lb/in.	2.0	4.5
Peel Adhesion,	kN/m	0.4	2.1
	lb/in.	2.5	12.0
Shear Adhesion, s		1.6	32.3

	No. 121	No. 122	No. 123
	(Polybutene/Polyethylene)		
LDPE (low melt index, 24.5 g/10 min)	40	—	—
LDPE (med. melt index, 155 g/10 min)	—	40	—
LDPE (high melt index, 2100 g/10 min)	—	—	40
Tackifier	40	40	40
Wax	20	20	—
Antioxidant	1	1	1
Indopol® Polybutene L-014	20	—	—
Indopol® Polybutene H-100	—	30	—
Indopol® Polybutene H-1500	—	—	20

Procedure:

In the laboratory, the components are blended with constant mixing at 204 C (400 F) under a nitrogen blanket until the mix is homogeneous.

Properties:

	No. 121	No. 122	No. 123
Visc. @ 149 C (300 F), cps	4700	1550	1300
Softening Pt., C	98	95	96
F	208	203	204
Tensile Strength, kPa	1215	1413	1275
psi	176.2	205.0	185.0
Elongation, %	40	44	124
Durometer Hardness	57	58	55
Needle Penetration, mm	2.3	2.2	1.4
T-Peel Adhesion:			
Natural Kraft, 60 lb, N/m width	252.8	313.0	350.2
oz/in. width	23.1	28.6	32.0
Polyethylene film, untreated,			
N/m width	167.4	233.1	164.2
oz/in. width	15.3	21.3	15.0

Aluminum foil, 0.0025 in.,

	N/m width	192.6	241.8	437.8
	oz/in. width	17.6	22.1	40.0

No. 124
(Polypropylene/Rosin)

Polytac R-500	50
Hydrogenated Rosin	30
Amoco® Polybutene H-300	20

Note:

The resulting product has excellent surface tack and a Brookfield visc. @ 34 F of 1300 cps.

No. 125

(Polypropylene/Rosin)

Polytac R-500	60
Tall Oil Rosin	40

Note:

The above mixture has excellent elongation. Softening point is 300 F. Brookfield visc. @ 340 F is 1100 cps.

No. 126

(Polypropylene/Rosin)

Polytac R-500	50
Tall Oil Rosin	34
Paraffin Wax 190/195	15

Note:

The resulting adhesive is tacky and stringy. The Brookfield visc. @ 340 F is 1000 cps; the softening point is 290 F.

No. 127

(Polypropylene/Resin)

Polytac R-500	70
Exxon 1315	20
Amoco® Polybutene H-300	10

Note:

The mixture has a softening point of 295 F and a Brookfield visc. @ 340 F of 3000 cps.

No. 128

(Polypropylene/Resin)

Polytac R-500	50
Foral® 85	30
Amoco® Polybutene H-300	20

Note:

The resulting product has a softening point of 295 F and a Brookfield visc. @ 350 F of 2000 cps. The resulting adhesive can be used as a permanent label adhesive.

No. 129

(Polypropylene/Resin)

Polytac R-500	60
Exxon 1315	40

Note:

The mixture results in superior bonding strength, with no delamination at 140 F after 24 h.

	No. 130	No. 131	No. 132	No. 133	No. 134
Ethyl Cellulose N-4	—	20	—	—	17
Ethyl Cellulose N-10	25	40	20	20	—
Staybelite® Ester 10	50	—	—	—	—
Nylon Powder	—	10	—	—	53
Abalyn	15	5	—	—	30
Paraffin Wax	5	—	5	5	—
Castor Wax	5	25	—	25	—
Staybelite® Resin	—	—	—	50	—
Pentalyn® H	—	—	50	—	—
Pentaphen No. 67	—	—	25	—	—

	No. 135	No. 136
	(Resin)	
Pale Crepe	50.0	50.0
Wingtack® 95	50.0	—
Polyterpene Resin (1010 C)	—	50.0
Tricresyl Phosphate	5.5	5.5
Wingstay L	0.5	0.5
Toluene	318.0	318.0

No. 137

(Resin)

Versalon 1193	79.0
FF-680	20.0
Sb_2O_3	5.0
Butyl Zimate®	1.0

Procedure:

Melt together with stirring the **Versalon** and the antioxidant. When molten, add the solid **FF-680** and Sb_2O_3 with stirring, and use directly or cast into appropriate shapes for later use.

Hot-Tack Heat-Seal Coating

Piccotex® LC	25.0
EVA (28% VAc, 6 MI)	30.0
Paraffin Wax (m.p. 155 F)	45.0
BHT	0.1

Hot-Melt Curtain Coating

Piccotex® LC	15.0
EVA (28% VAc, 6 MI)	15.0
Paraffin Wax (m.p. 155 F)	60.0
Microcrystalline Wax (m.p. 180 F)	10.0
BHT	0.1

Heat-Resistant, Hot-Melt Adhesive

Formula No. 1

(Ethylene Vinyl Acetate/Resin)

EVA-607	40.0
CKM-2400	30.0
Piccotex® LC	10.0
Shell 700	15.0
Multiwax 180-M	5.0
BHT	0.5

	No. 2	No. 3
	(Styrene Butadiene/Resin)	
Solprene 301	100	100
Zinc Oxide	40	40
Pentalyn® H	100	—
Piccolyte® Alpha 125	—	75
Mineral Oil	10	10
Sulfur	3	—

Butyl Eight	8	—
Amberol ST 137	—	12
Stannous Chloride	—	4
Antioxidant 2246	2	2
Cure in 300 F oven, min	2	5

Hot-Melt Adhesive for Polyester

(Ethylene Vinyl Acetate/Resin)

EVA-501	50
Dymerex	15
Staybelite® Ester 10	15
Multiwax 180-M	20

Hot-Melt Adhesive for Polystyrene

(Ethylene Vinyl Acetate)

EVA-505	45.0
CKM-2400	40.0
Multiwax 180-M	15.0
BHT	0.5

Hot-Melt Adhesive for Polypropylene

(Ethylene Acrylic/Resin)

DPDA-9169	40.0
Dymerex	25.0
Piccolyte® S-115	15.0
Multiwax 180-M	20.0
BHT	0.5

Hot-Melt Adhesive for Polyethylene

(Ethylene Acrylic/Resin)

EAA-9060	40.0
Piccolyte® A-115	35.0
CKM-2400	5.0
Multiwax 180-M	20.0
BHT	0.5

Pressure-Sensitive Adhesive with Hot-Shear Strength

(Isobutylene-Isoprene/Resin)

Enjay Butyl LM 430	67
Enjay Butyl 077	33
Schenectady ST 5010	10
Schenectady SP-1055	10
Zinc Resinate	5

Cure 15 min @ 350 F.

Solvent-Based, Hot-Melt Adhesive

Propionate (alcohol soluble)	16.0
SAIB	16.0
Kodaflex DOP	8.0
Tecsol C (95%)	60.0

Hot-Melt, Pressure-Sensitive Masscoat

Formula No. 1

(Rubber/Resin)

Kraton® 1102	33.0
Kraton® 1107	67.0
Super Sta-Tac®	60.0

Sta-Tac® 100	20.0
Betaprene® AC-130	40.0
Antioxidant 330	0.5
DLTDP	0.5

Properties:

Melt Visc. @ 350 F	20,000 cps
180° Peel Strength	3.8 pli
Shear Adhesion @ 150 F	No creep under 1 lb load

Note:

Provides high-temperature stability during application.

	No. 2	No. 3
	(Rubber/Resin)	
Kraton® 1107	100.0	75.0
Kraton® 1102	—	25.0
Foral® 105	120.0	120.0
Antioxidant 330	0.5	0.5
DLTDP	0.5	0.5

Properties:

Melt Visc. @ 350 F, cps	17,500	39,000
Sol'n. Visc. # 30%w, cps	90	95
180° Peel Strength*, pli	6	6
Shear Adhesion to Glass**, h	10	140
Shear Adhesion to Cardboard**, h	2.5	6

---

*   PSTC #1

** PSTC #7 modified with 1 × 1 in. area for glass, 2 × 0.5 in. area for cardboard, and 5.5 lb load

Note:

Provides high holding power.

Uses:

Heavy-duty strapping tapes.

Hot-Melt Sealants

Formula No. 1

(Butyl Rubber)

Butyl (normal)	45
UE 634-04	10
Wood Rosin	10
Mineral Filler or Carbon Black	25
Liquid Polybutene	10

No. 2

(Butyl Rubber/Polypropylene)

Bucar 5214	30.9
Nirez® V-2150	7.7
Nirez® V-2040	7.7
Super Sta-Tac® 80	3.8
Betaprene® AC-130	7.7
Elvax 460	7.7
Afax®-510	23.0
Sunpar 150	3.8
IT-3X	7.7

Procedure:

Using a kneader-extruder, heat equipment to 330–350 F; charge all **Bucar 5214** and **Nirez® V-2150/Nirez® V-2040**; mix.   Charge all **Super Sta-Tac® 80/Betaprene® AC-130**;   mix.   Charge **Elvax 460/Afax®-510**;   mix.   Charge **Sunpar 150** and mix.   Mix formulation until the system is homogeneous.

Chapter IV

PRESSURE-SENSITIVE ADHESIVES

Pressure-Sensitive Tape

Formula No. 1

(Rubber/Resin)

Natural Rubber	100.0
Super Sta-Tac® 80	67.0
FF-680 (micronized)	50.0
FLX-0018	10.0
Sb$_2$O$_3$	12.5
Antioxidant 2246	1.0
Solvent*	500.0

---

* Hexane or hexane/toluene mixture to make 20% rubber solution.

Procedure:

Mill crepe rubber to a Mooney viscosity of 72 (or equivalent). Dissolve rubber in solvent, followed by the tackifier resin and antioxidant. To the rubber/resin solution, disperse the solids, grinding to a fine size, if desired. Apply to the substrate material and allow to air dry.

No. 2

(Rubber/Resin)

Kraton® Rubber	100
Wingtack® 95	100
Irganox® 1010	2

No. 3

(Rubber/Resin)

Kraton® D 1111	100
Escorez® 1310	140
Shellflex® 371	10
Piccovar® AP-33	40
Antioxidant	2

Pressure-Sensitive Reinforced Strapping or Bundling Tape Adhesive

Formula No. 1

(Rubber/Resin)

Zonarez B-70	8.0
Reclaimed Rubber	18.0
Naphthenic Mineral Oil	4.0
Aluminum Hydrate	4.0
Antioxidant	0.2
Benzene	65.8

	No. 2	No. 3	No. 4	No. 5
	(Styrene Butadiene/Resin)			
Solprene 406	60	50	50	50
Solprene 1205	40	—	—	—
Solprene 300	—	50	50	—
Solprene 301	—	—	—	50
Pentalyn® H	125	75	125	75

No. 6

(Styrene Butadiene/Resin)

Ameripol 1011	100
Pentalyn® H	75
Antioxidant 2246	2
Toluene	1003

Pressure-Sensitive General-Purpose Mending Tape Adhesive

Formula No. 1

(Rubber/Resin)

Zonarez B-85	60.0
Butyl Rubber (25% isoprene)	70.0
Polybutene (500 m.w.)	30.0
Polyisobutylene	30.0
Zinc Oxide	30.0
Polyparanitrobenzene	0.8
Heptane	800.0

No. 2

(Rubber)

Natural Rubber	100
Ester Gum	50–100
Lanolin	0–50
Antioxidant	1–2
Zinc Oxide	0–100
Solvent	as required for a 25–30% sol'n.

Pressure-Sensitive Low-Tack Protective Tape Adhesive for Polished Metal

(Rubber/Resin)

Zonarez B-85	10.0
Butyl Rubber (25% isoprene)	100.0
Paraffinic Oil	10.0
Hydrated Alumina	12.2
Polyparanitrobenzene	0.8
Heptane	600.0

Pressure-Sensitive Transparent Tape

(SBR/Resin)

Zonarez B-85	50
Natural Crepe Rubber	50
SBR (28% styrene)	50
Antioxidant	1
Heptane	600

Pressure-Sensitive Transparent-Film Paper Tape Adhesive

(Styrene Butadiene/Isobutylene-Isoprene)

Enjay Butyl HT 10-66	50
SBR 1011	50
Maglite K	1
Schenectady SP-567	30
Aroclor 1254	20

Pressure-Sensitive Electrical Tape

Formula No. 1

(Rubber/Resin)

Zonarez B-85	50
Ethylene Propylene Rubber	300
Butyl Rubber (1.0 mol isoprene)	100
Polyethylene (21,000 m.w.)	75
Paraffinic Oil	75
Diatomaceous Earth	250
Toluene	2000

No. 2

(Rubber/Resin)

Zonarez B-115	4.0
Zonarez B-70	48.0
Zonester® 85	15.0
Pale Crepe Rubber	60.0
SBR (12% styrene)	40.0
Reclaimed Rubber	5.0
Reactive Phenol Formaldehyde Resin	4.0
Triethylene Glycol Ester of Rosin	1.0
Alkyl Phenol Sulfide Curing Agent	2.3
Heptane	550.0

Note:
This formula is for low-temperature applications.

Pressure-Sensitive Low-Temperature
Tape Adhesives

	Formula No. 1	No. 2	No. 3	No. 4
	——— *(Rubber/Resin)* ———			
Kraton® D-1107	100	100	100	100
Escorez® 1310	80	60	90	—
Wingtack® 95	—	—	—	80

Wingtack® 10	—	20	40	—
Shellflex® 371	—	—	—	20
Antioxidant	2–5	2–5	2–5	2–5

Properties:

Tg (calculated), C	–22	–29	–22	–23
Dead Weight Tack (stainless steel probe), –18 C, g	230	340	400	310
Rolling Ball Tack, cm	1.4	1.0	1.1	1.0
Polyken® Probe Tack at room temperature, kg	1.1	0.65	0.95	0.72
Loop Tack, N/m	1060	820	1060	900
180° Peel Adhesion, N/m	670	550	630	470
Holding Power, min	> 6000	> 6000	> 6000	> 6000
Melt Visc. @ 177 C, Pa·s	300	215	60	130

Pressure-Sensitive Reinforced Packaging Tape

(Rubber)

Natural Rubber	100
Ester Gum	175
Lanolin	25
Antioxidant	1
Zinc Oxide	50
Solvent	400

Pressure-Sensitive Surgical Tape

(Rubber)

Natural Rubber	100
Ester Gum	100
Lanolin	25
Antioxidant	1
Zinc Oxide	100
Solvent	200

Pressure-Sensitive Sealing Tape

Formula No. 1

(Rubber)

Hycar® 4054	100
Hydral 710	200
N-550 Black	40
TCP	50

No. 2

(Nondrying, Polybutene)

Amoco® H-300	24.01
Amorphous Polypropylene Homopolymer	5.05
Butyl Rubber	1.72
Clay	16.37
Calcium Carbonate	44.05
Diatomaceous Silica	4.00
Cotton Fiber	4.80

Procedure:

This is compounded in a sigma-blade mixer. Amorphous polypropylene is premixed with twice its weight of polybutene to facilitate complete dispersion. Additions to the mixer are made in the order of listing in the formula. The entire mass is then mixed for 1 h after the last addition.

	No. 3	No. 4	No. 5
	(Polybutene/Rubber)		
Amoco® H-300	18.90	19.90	32.0
Polyisobutylene Rubber	18.90	19.90	—
Butyl Rubber	5.25	—	8.0
Calcium Carbonate	35.05	35.05	35.0
Platelet Talc	6.50	7.50	—
Attapulgas Clay	11.00	12.25	16.0
Diatomaceous Silica	2.40	3.40	5.0
Titanium Dioxide	2.00	2.00	4.0

Procedure:
In the laboratory the rubbers and polybutene are mixed in a dual-arm sigma-blade mixer for approximately 20 min. Dry ingredients are incrementally added in the order listed while mixing. The sealant is dumped, rested 24 h, and extruded.

Pressure-Sensitive Polyethylene-Backed Tape

(Isobutylene/Butyl Rubber)

Enjay 218	16.5
Polyisobutylene (1200 m.w.)	17.5
Hexane	66.0

Note:
This formulation is resistant to aging, oxidation, and chemical reaction.

Pressure-Sensitive Masking Tapes Adhesives

	Formula No. 1	No. 2	No. 3	No. 4	No. 5
			(Isobutylene/Resin)		
Vistanex L-100	27	27	36	35	35
Polybutene 128	14	13	10	18	22
Oletac 100	13	14	—	—	—
Staybelite® Ester 3	—	13	18	—	17
Staybelite® Ester 10	—	—	9	—	—
Hercolyn®	—	—	—	—	13
Abitol®	—	—	—	21	—
Piccolyte® S-115	13	13	—	—	—
Mineral Oil	20	20	27	26	26

No. 6

(Isobutylene/Resin)

Polyisobutylene (M_{vis} 100,000)	35
Polybutene 32 or 128	21
Beta Pinene Polymer Resin (m.p. 100 C)	20
Dihydromethyl Abietate	14
Monobutyl Ether of Ethylene Glycol Stearate	10
Naphtha	to desired consistency

Note:

This gives a clear pressure-sensitive formulation claimed to be resistant to effects of air and sunlight. It should, therefore, be particularly suitable as an adhesive for masking tapes having a transparent film such as cellulose acetate as the support.

No. 7

(Styrene Butadiene/Resin)

Ameripol 1006	100.0
Foral® 105	50.9
Gulf 534	37.5
Antioxidant 2246	2.0
Toluene	861.0

No. 8

(Styrene Isoprene/Resin)

Zonarez B-115	10
Styrene/Isoprene Block Copolymer	100
Zinc Resinate	7
Reactive Phenol Formaldehyde Resin	7
Antioxidant	3
Toluene	175

Note:

This is a high-performance cured masking tape adhesive.

Pressure-Sensitive, Hot-Melt, Heat-Resistant Masking Tape Adhesive

(Rubber/Resin)

Zonarez B-115	100.0
Smoked Sheet Rubber	75.0
SBR (24% styrene)	25.0
Zinc Oxide	50.0
Alkyl Phenol Sulfide Curing Agent*	8.0
Mineral Oil	3.0
Antioxidant	1.0
Heptane	350.0

---

* Cure at 300–350 F.

Pressure-Sensitive Vinyl Tape Adhesive

(Styrene Butadiene/Resin)

Ameripol 4505	100
Foral® 85	75
Antioxidant 2246	2
Toluene	1003

Pressure-Sensitive Pigmented Paper Tape Adhesive

(Styrene Butadiene/Isobutylene-Isoprene)

Enjay Butyl HT 10-66	50
SBR 1011	50
Whiting	100
Maglite K	1
Schenectady SP-567	20
Indopol® H-1900	40

Cloth Mending Tape Adhesive

(Isoprene/Resin)

Ameripol SN 600	100
Staybelite® Ester 10	85
Antioxidant 2246	2
Toluene	1059

Pressure-Sensitive Adhesive for Tapes with Soft Backing Material

Formula No. 1

(Acrylic Resin)

Acronal 85 D (pH 7–8)	60
Acronal 4 D (pH 7–8)	20
Lutonal I 65 D	20

No. 2

(Acrylic Resin)

Acronal 85 D	80.00
Acronal 30 D	20.00
Plastilit 3060	0.75

No. 3

(Acrylic Resin)

Acronal 80 D (pH 7–8)	80
Lutonal I 65 D	20

Note:

After drying, the coat weight should lie at 25–30 g/m². The above three adhesives have good adhesion to plasticized PVC films which are recommended for electrical insulating tapes. Good adhesion to such films is obtained even without the application of a primer.

There are no interactions between polyethylene backings and the adhesive. In order to obtain adequate anchorage, it is necessary to subject the material to a Corona pretreatment.

No. 4

(Acrylic Resin)

Acronal 85 D	90
Acronal 7 D	10

Pressure-Sensitive Adhesive for Tapes with Hard Backing Material

Formula No. 1

(Acrylic Resin)

Acronal 80 D (pH 7–8)	80
Lutonal I 65 D	20

No. 2

(Acrylic Resin)

Acronal 85 D	90
Acronal 7 D	10

No. 3

(Acrylic Resin)

Acronal V 205	90
Acronal 7 D	10

No. 4

(Acrylic Resin)

Acronal V 205	50
Acronal 85 D	50

Note:

These adhesives can be applied by various methods. The above-mentioned formulations must be suited to the processing requirements by modification (e.g., by altering the proportions of the constituents, and adding wetting and thickening agents).

Polypropylene must be pretreated electrically in order to obtain adequate adhesion.

No. 5

(Acrylic Resin)

Acronal 500L (ca. 40% in ethyl acetate)	80
Acronal 4L (ca. 50% in ethyl acetate)	10
Carbigen® K90 (50% sol'n. in toluene)	10

No. 6

(Acrylic Resin)

Acronal 4L (ca. 25% in acetone/mineral spirit)	60
Acronal 500L (ca. 40% in ethyl acetate)	25
Staybelite® Ester 10 (50% sol'n. in toluene)	15

Properties:

The coat weight after drying should lie at 20–30 g/m².

Pressure-Sensitive Adhesive for Double-Faced Tape

Formula No. 1

(Acrylic Resin)

Acronal 80 D	60
Hercolyn® 1151 (60% sol'n. in toluene)	40

No. 2

(Acrylic Resin)

Acronal 85 D	60
Hercolyn® 1151 (60% sol'n. in toluene)	40

Note:

Double-faced adhesive articles are supplied in the form of tapes, films, and cuttings with and without carrier sandwiched between silicone release paper. Tapes are used for fixing edges (carpet tapes), and carrierless adhesive films for assembling work. Films are used for laminating large-size prints.

The coat weight of double-faced articles lies at 60–100 g/m² per side.

No. 3

(Acrylic Resin)

Acronal 30 D	95
Acronal 4 D	5

No. 4

(Acrylic Resin)

Acronal 35 D	50
Acronal 50 D	50

No. 5

(Acrylic Resin)

Acronal 3 L (ca. 25% in acetone/mineral spirit)	100.0
Desmodur® L (75% in ethyl acetate)	0.5

Properties:

The coat weight should lie at 5–10 g/m².

Pressure-Sensitive Decorative Film Adhesive

Formula No. 1

(Acrylic Resin)

Acronal 80 D (pH 7–8)	70
Lutonal I 65 D	30

No. 2

(Acrylic Resin)

Lutonal M 40 (ca. 50% in water)	40.0
Acronal V 302	60.0
Water	20.0
Toluene	2.5

Procedure:

First dilute **Lutonal M 40** (ca. 50% in water) with water, and then stir in **Acronal V 302**. The toluene in the mix retards the precipitation of **Lutonal M 40** at temperatures above approx. 28 C.

No. 3

(Acrylic Resin)

Acronal 80 D	50
Acronal 880 D	50

No. 4

(Acrylic Resin)

Acronal 85 D (pH 7–8)	80
Lutonal I 65 D	20

Properties:

The coat weight after drying should lie at 20–30 g/m^2.

No. 5

(Acrylic Resin)

Acronal 4 L (ca. 40% in acetone/mineral spirits)	60
Acronal 500 L (ca. 40% in ethyl acetate)	25
Staybelite® Ester 10 (50% sol'n. in toluene)	15

No. 6

(Acrylic Resin)

Acronal 500 L (ca. 40% in ethyl acetate)	80
Acronal 4 L (ca. 50% in ethyl acetate)	10
Carbigen® K 90 (50% sol'n. in toluene)	10

Properties:
The coat weight after drying should lie at 20–30 g/m^2.

Note:
The **Acronal L** types are used for such adhesives. The initial tack of the adhesive can be increased to the required level by adding resins. Pressure-Sensitive adhesives based on **Acronal L** types have a tendency to demix and should, therefore, be thoroughly stirred before application on the machine.

Special care must be taken that no interactions can take place between the backing material and the adhesive (e.g., by plasticizer migration). The films available on the market may differ in their behavior. For this reason, it is necessary to test the adhesive together with the film to be used.

Pressure-Sensitive, Plasticized Decorative Film Adhesives

Formula No. 1

(Acrylic Resin/Vinyl Ether)

Acronal 80 D	80
Lutonal I 65 D	20

No. 2

(Acrylic Resin/Vinyl Ether)

Acronal 80 D	70
Lutonal M 40 (ca. 50% in water)	30
Ethyl Acetate	5

Note:

At temperatures above 28 C, mixtures of **Lutonal M 40** and polymer dispersions are unstable on account of the solubility inversion of the vinyl methyl ether polymers.   A slightly gritty to completely coagulated mixture is formed, depending on the temperature and duration of exposure. These mixtures can be rendered stable even at temperatures above the precipitation point of **Lutonal M 40** by adding small quantities of solvent, if necessary, in combination with an emulsifier.

Pressure-Sensitive Laminating-Film Adhesives

Formula No. 1

(Acrylic Resin)

Acronal 500 D	70
Acronal 4 D	30

No. 2

(Acrylic Resin)

Acronal 500 D	80
Acronal 80 D	20

No. 3

(Acrylic Resin)

Acronal 500 D	90
Acronal 85 D	10

Properties:

The coat weight of the pressure-sensitive adhesive for laminating films should lie at 40–50 g/m² after drying.

Adhesives based on **Acronal 4 D** can be smoothly peeled off. Adhesives based on **Acronal 80 D** and **85 D** have a more aggressive tack.

Note:

Pressure-sensitive laminating films are used for protecting books, maps, and documents. The film for maps and documents should have particularly high dimensional stability. Plasticized PVC film is normally used as backing material for these applications. In some cases a "greyish cast" may form on books with a coarse linen cover. This effect should be reduced as much as possible by applying a high coat weight and rubbing the film tightly on the substrate. The initial tack must be adjusted in such a manner that corrrections can be made during the laminating process without damaging the substrate.

Pressure-Sensitive Rubber-Hose Adhesive

(Isobutylene-Isoprene)

Enjay Butyl HT 10-66	100
SRF	60
MT	90
Flexon 845	5
Paraffin Wax (133 F m.p.)	5
Stearic Acid	3
Antioxidant 2246	2
Zinc Oxide	5
Ledate	2

Procedure:

Steam cure 20 min @ 320 F.

Pressure-Sensitive Mass Coats

	Formula No. 1	No. 2
	(Rubber/Resin)	
Kraton® 1101	100.0	100.0
Foral® 105	200.0	200.0
Paraffinic Oil	—	10.0
Antioxidant 330	0.3	0.3
DLTDP	0.3	0.3

Properties:

		Formula No. 1	No. 2
180° Peel Strength, pli		3.6	5.5
Shear Adhesion		> 1 month[1]	> 1 week[2]
Oil Bleed Out @ 158 F,	24 h	—	none
	96 h	—	none

[1] PSTC #2 modified with 1.0 in.[2] area, 500 g lead
[2] PSTC #2 modified with 0.25 in.[2] area, 1000 g lead

Note:

Holding power may be increased further by reducing the resin level to 100 parts. Oil concentrations up to 20 phr may be tolerated without bleedout. Further addition of oil will increase tack at the expense of peel strength and shear adhesion.

Uses:

Where high holding power is required and aggressive tack is not.

	No. 3	No. 4
	(Rubber/Resin)	
Kraton® 1107	100.0	100.0
Foral® 105	125.0	125.0
Cumar® LX-509	—	15.0
Antioxidant 330	0.3	0.3
DLTDP	0.3	0.3

Properties:

Tensile Strength @ 122 F, psi	450	750
Shear Adhesion @ 176 F,		
PSTC #2, $^1/_2$ in.^{2}, 500 g lead, min	10	16

Note:

The addition of a high melting point resin (150 C or higher) which associates with the styrene diamines, such as **Cumar® LX-509**, tends to increase the adhesive strength at temperatures in the range of 140–175 F.

Uses:

Where aggressive tack is required.

No. 5

(Rubber/Resin)

Kraton® 1107	100.0
Piccolyte® A-125	45.0
Shellwax 300	0–10.0
Antioxidant 330	0.5
DLTDP	0.5

Properties:

The resulting compound has high tack, good holding power, and high melt viscosity when no wax is used. Adding wax will reduce tack, holding power, and melt viscosity.

Note:

Substituting **Kraton® 1102** for **1107** will increase cohesive strength at the expense of tack. Substituting **Piccopale® 200, Picco® Phenolic LTP-135**, or **Piccotex® 75** will increase cohesive strength at the expense of tack. A paraffinic oil may be used instead of paraffinic wax.

Uses:

Where aggressive tack but limited holding power is required as in self-stick carpet tiles.

	No. 6	No. 7
	(Rubber/Resin)	
Kraton® 1107	100.0	100.0
Foral® 105	125.0	125.0
Cumar® LX-509	—	15.0
Antioxidant 330	0.3	0.3
DLTDP	0.3	0.3
Properties:		
Tensile Strength @ 122 F, psi	450	750
Shear Adhesion @ 176 F,		
PSTC #2, $^1/_2$ in.^{2}, 500 g lead, min	10	16

Note:

The addition of a high melting point resin (150 C or higher) which associates with the styrene diamine, such as **Cumar® LX-509**, tends to increase the adhesive strength at temperatures in the range of 140–175 F.

Uses:

Where aggressive tack is required.

	No. 8	No. 9	No. 10	No. 11
		(Rubber/Resin)		
Kraton® 1101	—	—	—	25.0
Kraton® 1107	—	25.0	50.0	—
Natural Rubber (ML55)	100.0	75.0	50.0	75.0
Foral® 105	100.0	100.0	100.0	100.0
Antioxidant 330	0.3	0.3	0.3	0.3
DLTDP	0.3	0.3	0.3	0.3
Properties:				
Sol'n. Visc., cps	7500	3000	2000	3000
180° Peel Strength[1], pli	1.5	1.5	1.5	1.5
Shear Adhesion to Glass[2], h	1	1.5	7	8

---

[1] PSTC #1 1 in. wide cellophane tape applied to steel with 4.5 lb roller.
[2] PSTC #7 0.5 in.^{2} overlap with 11 lb load.

Note:

Films of **Kraton®** 1107/NR are somewhat clearer than films of **Kraton 1101/NR**. Similar improvement may be expected in other types of NR adhesives through addition of **Kraton®** rubber.

Uses:

Where high-shear adhesion and lower solution viscosity are required in NR-based adhesives.

	No. 12	No. 13	No. 14	No. 15
		(Polyisoprene/Resin)		
Kraton® 1107	—	25	50	75
Natsyn 400	100	75	50	25
Wingtack® 95	100	100	100	100
HiSil 233	50	50	50	50
Atomite®	50	50	50	50
Wingstay L	1	1	1	1
Properties:				
180° Peel Strength, pli	2.6	2.2	2.0	1.9
Shear Adhesion, h	625	430		
weeks,			> 4	> 4
Rolling Ball Tack, in.	3	10	15	> 15

Note:

Wingtack® 95 is a synthetic resin which works well with **Kraton® 1107** as well as with polyisoprene or NR.

Uses:

Where improved shear adhesion is required in a polyisoprene-based adhesive.

	No. 16	No. 17
	(Polybutene/Resin)	
Elastomer (solids)	100	100
Foral® 85	200	—
Amoco® Polybutene	0–150	0–150

Piccolyte® S-115		—	70
Solvent		20% toluene	toluene
		80% naphtha	—

Pressure-Sensitive Mass Coats
for Permanent Adhesives

	Formula No. 1	No. 2	No. 3	No. 4	No. 5
	(Styrene Butadiene/Resin)				
Solprene 301	100	100	100	—	
Solprene 375	—	—	—	—	137.5
Pentalyn® H	100	125	—	—	—
Piccolyte® Alpha 100	—	—	100	125	—
Piccolyte® Alpha 115	—	—	—	—	100.0

Pressure-Sensitive Mass Coats
for Temporary Adhesives

	Formula No. 1	No. 2	No. 3
	(Styrene Butadiene/Resin)		
Solprene 375	137.5	137.5	137.5
Pentalyn® H	50	—	—
Staybelite® Ester 10	—	50.0	—
Piccolyte® Alpha 100	—	—	50.0

Pressure-Sensitive, High-Strength Adhesive

Formula No. 1

(Rubber/Resin)

Kraton® 1107	50.0
Kraton® 1101	50.0
Super Sta-Tac® 100	90.0
Super Beckacite® 2000	40.0
Antioxidant	1.0
Toluene	150.0
Hexane	150.0

Procedure:

Predisperse resins and solvents. When resin is in and free from lumps, add the antioxidant and **Kraton**® Polymer.

Properties: (System coated 2 mils dry on Mylar.)

Solids	43.5%
180° Peel	6.0 lb
Quick Stick	2475 g
Room Temp. Shear[1]	No failure up to 72 h
180 F Shear[2]	Failure @ 105 min

[1] $1/_2 \times 1/_2$-in. sample with 2.5 kg load
[2] Load: 100 g/in.[2]

Note:

System can be further diluted to require solids/viscosity with aerosol propellants.

	No. 2	No. 3	No. 4	No. 5
	(Styrene Butadiene/Resin)			
Solprene 406	100	100	100	50
Solprene 1205	—	—	50	50
Pentalyn® **H**	100	100	100	—
Sunpar 110	25	25	—	—

Nonskid Adhesive

(Styrene Butadiene/Resin)

Solprene 414P	100.0
Amoco® **Resin 18-210**	30.0
Zinc Oxide	20.0
Flexon 766	60.0
Nevtac 80	5.0
Antioxidant	3.0
UV Inhibitor	0.1

Pressure-Sensitive, General-Purpose Adhesives

	Formula No. 1	No. 2
	(Rubber/Resin)	
Kraton® D-1107	100	100
Super Nirez™ 5100	100	—
Super Nirez™ 5120	—	100
Irganox® 1010	1	1
Hexane (90%)	271	271
Toluene (10%)	30	30

Properties: (Samples coated 1.0 dry mils on 2.0 mil Mylar type-"S")

Rolling Ball Tack, in.	> 12	> 12
Quick Stick (loop test), $g/in.^2$	1996	2220
Polyken® Probe Tack,		
(spread 0.1 cm/s, dwell-instant), g	1143	1201
180° Peel Adhesion		
(to stainless steel), g/in. of width	2200	2100
0° Shear Adhesion @ 77 F		
(2 $kg/in.^2$ load), days	> 7	> 7

No. 3

(Rubber/Resin)

Natural Rubber Pale Crepe #1	100
Super Nirez™ 5100	70
Antioxidant	1
Hexane (90%)/Toluene (10%)	to suit visc. requirements

Properties: (Sample coated 1.5 mils on tape stock)

Rolling Ball Tack	$^1/_2$–$^3/_4$ in.
Peel Adhesion	1400+ g/in. of width
Shear Adhesion (1 psi load @ 60 C)	40 min

	No. 4	No. 5	No. 6	No. 7	No. 8
		(Rubber/Resin)			
Kraton® D 1111	100.0	100.0	100.0	100.0	100.0
Escorez® 1310	100.0	140.0	120.0	100.0	100.0
Cumar® LX-509	—	—	20.0	40.0	—
Piccovar® AP-25	—	—	—	—	40.0
Shellflex® 371	10.0	10.0	10.0	10.0	10.0
Irganox® 1010	1.0	1.0	1.0	1.0	1.0
Toluene	140.7	167.3	167.3	167.3	167.3

No. 9

(Rubber/Resin)

Kraton® 1101	100.0
Foral® 85	200.0
Antioxidant 330	1.0
Shellflex® 371-N	20.0

Note:
The above dissolves slowly in toluene or cyclohexane.

	No. 10	No. 11	No. 12
	(Rubber/Resin)		
Kraton® D 1111	100	100	100
Escorez® 1310	140	140	140
Piccovar® AP-25	—	40	35
Cumar® LX-509	—	—	35
Shellflex® 371	10	10	40
Antioxidant	1	1	1

	No. 13	No. 14	No. 15
	(Rubber/Resin)		
Kraton® D-1107	100	100	100
Wingtack® 95	80	—	—
Wingtack® 76	—	80	—
Wingtack® 10	—	—	80
Properties:			
T_g (calculated), C	−16	−22	−45
Dead Weight Tack			
@ −12 C, g	150	180	—
@ −18 C, g	6	190	290
@ −29 C, g	—	—	310

	No. 16	No. 17	No. 18
	(Rubber/Resin)		
Kraton® D-1107	100	100	100
Wingtack® 95	125	—	125
Wingtack® 76	—	125	—
Shellflex® 371	25	25	—
Indopol® L50	—	—	50
Properties:			
T_g (calculated), C	−12	−19	−23
Dead Weight Tack			
@ −12 C, g	35	160	225
@ −18 C, g	0	60	180

No. 19

(Rubber/Resin)

Milled Rubber, Polybutene, Vistanex, and Butyl Rubber	40–70
Piccolyte® S-115, S-85, or **S-100**	20–30
White Oil	10–30
Antioxidant	as required

Note:

Pigments such as zinc oxide, and fillers such as calcium carbonate, clay, and aluminum hydrate are added often in amounts up to the weight of rubber-like materials employed.

Solvents when employed are usually rather volatile and hexane is often used, although light naphtha and mineral spirits are also used. Aromatic solvents, such as toluene or xylene, may be used as required.

	No. 20	No. 21	No. 22
	(Rubber/Resin)		
Natural Rubber	56	—	—
Natsyn 2200 Rubber	—	46	—
Vistanex L-100	—	—	38
Easto-Rez	38	42	20
Abitol®	6	12	—
Polybutene H-100	—	—	30
Piccolyte® S-25	—	—	12
Solvent System:			
Toluene	60	60	60
Heptane	40	40	40

No. 23

(Latex/Resin)

Tylac® 97-453	100.000
Staybelite® Ester 10-50W (acid stable)	100.00
Vulcarite 775	0.50
Tychem® 01834	0.20
Ammonia	to pH 9.5

Dry 10 min @ 70 C.

	No. 24	No. 25	No. 26
	(Styrene Butadiene/Resin)		
Plioflex PFR 1011	100.0	—	—
SBR 1011 (A)	—	100.0	—
SBR 1011 (B)	—	—	100.0
Wingtack® 76	90.0	90.0	90.0
Antioxidant	1.0	1.0	1.0
Toluene	22.1	22.1	22.1
Hexane	423.4	423.4	423.4
Properties:			
Total Solids, %	30.9	30.6	30.8

No. 27

(Styrene Butadiene/Resin)

Solprene 414P	100
Neville LX-685 (125)	50
Zonarez 7100	50

No. 28

(Styrene Butadiene/Latex/Resin)

Natural Latex	50
Carboxylated SBR	50
Piccopale® 85	75
Pentalyn® GH	25

	No. 29	No. 30
	(Styrene Butadiene/Resin)	
Solprene 414P	100	60
Solprene 1205 C	—	40
Foral® 85	110	100
Antioxidant	5	5

	No. 31	No. 32	No. 33	No. 34
	(Styrene Butadiene/Resin)			
Solprene 1205C	50	50	60	70
Solprene 411	50	50	40	30
Pentalyn® H	100	—	—	—
Super Sta-Tac® 80	—	100	100	100
Antioxidant	2	2	2	2
Toluene	465	465	465	465

Properties (of 1 mil films on 1 mil Mylar):

Tack[1], g	560	845	860	990
180° Peel[2] lb/in.	4.4	4.0	6.6	7.6
kN/m	0.77	0.70	1.15	1.33
90° Holding Test[3], 200 g/in.	———— no creep————			
180° Holding Test[3], 2000 g/in.	———— no creep————			

---

[1] **Polyken®** Probe Tack Tester: 100 g/cm^2, 1 s/ 1 cm/s
[2] PSTC #1
[3] After 24 h

No. 35

(Styrene Butadiene)

Purelast 166	45
Rosin or Glycerol Ester of Rosin	10
IBOA	40
EHA	0–10
Photoinitiator	5

Note:
    Addition of only a very small amount of polyfunctional monomer will eliminate tack.

	No. 36	No. 37
	(Styrene Butadiene/Gum)	
Standard Stock*	121.00	—
Plioflex 1551	—	100.00
Hydrogum 400	100.00	100.00
Antioxidant 330	—	1.00
Hexane	330.00	300.00
Toluene	80.00	70.00
Properties:		
Total Solids, %	35.6	35.7
Visc., cps	1672	3868
Shear Adhesion ($1/_2$ in. @ 1000 g), min	168	234
180° Peel Adhesion, lb/in.	2.7	3.6
Aged Shear Adhesion ($1/_2$ in. @ 1000 g, 1 week @ 158 F), min	106	303
Aged 180° Peel Adhesion (1 week @ 158 F) lb/in.	2.55	3.55

---

* Stock Formulation:

Plioflex 1551	100.00
Aluminum Hydrate	20.00
Antioxidant 330	1.00

Milling Time	15 min
Milling Temp.	130 F
Mooney Visc. (ML 4 @ 212 F)	47

No. 38

Polyisobutylene (100,000 m.w.)	10
Atactic Polypropylene (30,000 m.w.)	10
Polybutene 32	5
Mineral Oil	5
Petroleum Naphtha (low boiling)	55
Toluene	15

	No. 39	No. 40	No. 41
	(Polyisobutylene/Butyl Rubber)		
Butyl Rubber	10	15	30
Polyisobutylene	30	45	40
Carbon Black	30	15	30
Silene EF	30	—	—
Clay	—	25	—

No. 42

(Resin)

Pale Crepe	100
SP-559	350
ST-5115	35
B-Naphtha Amine	1

No. 43

(Rubber/Resin)

Natural Mechanical Reclaim	200
Limed Rosin	50
SP-559	100
Zinc Oxide	50
Gasoline	240
Alcohol	160

Pressure-Sensitive Low-Cost Adhesive

(Resin)

Ethyl Cellulose N50	8
Staybelite® or Staybelite® Ester 10	51
Hercolyn® D	40
BHT	1
Solvent	to suit

Note:
This adhesive gives good low-temperature flexibility and bond strength.

Pressure-Sensitive Glass-Cloth Adhesive

(Butyl Isoprene)

Enjay Butyl LM 430	67
Enjay Butyl 365	33
Zinc Oxide	25
Nytal 300	100
Schenectady SP 567	15
Schenectady SP 1055	15

Pressure-Sensitive Elastomer-Solution Adhesives

	Formula No. 1	No. 2	No. 3
		(Polybutene/Resin)	
Elastomer Solids:			
Synthetic Natural Rubber	100	—	—
Polyisobutylene	—	100	—
SBR	—	—	100
Tackifier Resin	70	70	70
Indopol® H-100	75	—	—
Indopol® H-300	—	100	—
Indopol® H-1500	—	—	75
Toluene		to desired visc.	

Procedure:

In the laboratory all materials were added to toluene and mixed until completely dissolved.

Properties:

Rolling Ball Tack,	cm	1.9	> 30	> 30
	in.	0.7	> 12	> 12
Quick Stick Adhesion	kN/m	0.4	0.7	0.2
	lb/in.	2.0	4.2	1.1
Peel Adhesion	kN/m	0.5	1.1	0.5
	lb/in.	2.6	6.0	2.7
Shear Adhesion, s		175	460	930

Pressure-Sensitive, Water-Dispersed Adhesive

(Polybutene/Resin)

Thermoplastic Elastomer	100
Aromatic Solvent	75
Indopol® L-14	75
Amoco® Resin 18-210	40
Emulsifiers	8
Antioxidant	2
Water	90
Tackifier Sol'n.:	
Resin	285
Hexane	95

Procedure:

Heat **Resin 18-210** in solvent at 80 C (176 F) until dissolved. Add polybutene preheated to 80 C (176 F) and stir. Pour solution over elastomer and heat 15 min in an oven at 99 C (210 F). Mix and return to oven. Alternately heat and mix until homogeneous. Mix on a high-speed stirrer at 2000 rpm. Gradually add hot [99 C (210 F)] emulsifiers and antioxidant. Add hot [99 C (210 F)] water slowly until inversion occurs. Add remaining water. Mix for 10 min. Blend in tackifier.

Properties:

Rolling Ball Tack*	6.0 cm
	2.4 in.
Quick Stick Adhesion*	1.0 kN/m
	5.5 lb/in.
Peel Adhesion*	1.5 kN/m
	8.8 lb/in.
Emulsion Stability (shelf life, days to separation at room temp.)	
100% undiluted	60+
10% dilution	10+
Freeze-Thaw Cycles** Passed	10

---

* Using Grade A airplane fabric as the substrate.
** 8 h @ –12 C (10 F), 16 h @ R.T.

Chapter V

SPECIALTY ADHESIVES

FOAM ADHESIVES

Strong-Foam Carpet Tile Adhesive

(Acrylic)

Rhoplex® N-560 or N-619	71.1
Rhoplex® B-85	25.6
Acrysol® AS#-60/Water (1/1) or	3.0
Acrysol® GS @ 5.0	
Ammonium Hydroxide	0.3 to pH 8

Properties:
Visc. #4/60 rpm	20,000 cps
Knife Coating	0.5 oz/yd^2
Dried	3 min/240 F

Weak-Foam Carpet Tile Adhesive

(Acrylic)

Rhoplex® LC-67	100.0
Vinol® 540 (12%)	1.5–2.0

Properties:
Visc. #4/20 rpm	3000–4000 cps
Direct Roll Coating	1.5–2.0 g/ft^2

Fabric-Backed, Vinyl-to-SBR Foam Adhesive

	Formula No. 1	No. 2	No. 3
		(Dry, Latex)	
Hycar® 2679X6	100.0	—	—
Hycar® 1872X6	—	100.0	—
Geon® 450X20	—	—	100.0
Sodium Polyacrylate	—	1.0	—
Acrysol® ASE 60	0.5	—	—
Methocel® MC (4000 std.)	—	—	0.8
Ammonium Hydroxide	adjust to pH 6.3	—	—

Vinyl Foam-to-Metal Adhesive

(Polyvinyl Chloride Resin)

Tenneco 1730	100
Kodaflex PA-5	70
Kodaflex HS-3	15
Epoxidized Soybean Oil	5
Thermolite 25	2
Dibasic Lead Phthalate	6
Talc	10
X-970	8
Azodicarbonamide	amount dependent on required foam density

Fusion conditions: 10 min, 177 C (350 F)

Flexible Polyurethane Foam Assembly Adhesive

(Rubber/Resin)

Kraton® 1101	100.0
Foral® 105	50.0
Pentalyn® C	100.0
Antioxidant 330	0.3
DLTDP	0.3

Properties:

When bonding polyurethane foam to itself, the adhesive had good flexibility following 72 h, aging in an air oven at 212 F.

When bonding polyurethane foam to wood, the adhesive has good creep resistance as measured at 158 F for a period of 1 h. The adhesive also has good "quick-grab."

Note:

Kraton® 1102 may be used in place of **Kraton®** 1101 to decrease solution viscosity at the expense of high-temperature performance.

Uses:

Sprayable adhesive for flexible polyurethane foam.

Polyurethane Foam-to-Flannel Adhesive

(Rubber/Resin)

Hycar® 2671 (52%)	100
Titanium Dioxide (50% dispersion)	3
Carbopol® 934 (5%)	1
Ammonium Hydroxide (10%)	to raise pH to 8

Procedure:

Prepare a solution of **Carbopol®** 934 dispersed at 5% by weight in water. Neutralize the acid **Carbopol®** resin with ammonium hydroxide to develop a thickened mucilage. Stir **Hycar®** 2671 into the mucilage. Thicken the compound by further neutralization with 10% ammonium hydroxide.

Stretch the flannel cloth, with backside up, over a glass plate. Apply a 15-mil wet layer of the thickened latex over the entire area. Cure for 10 min at 250 F. Whip the remaining thickened latex in a Hobart mixer equipped with a wire beater. After 5–8 min of high-speed agitation, the emulsion will expand to approximately 500% of original volume.

Pour the whipped latex over the cured seal coat of the flannel. Set the doctor blade to approximately 25 mils above the fabric.

Place the polyurethane foam quickly over the latex and hold fast for 10 s with a glass plate. Cure the assembly 10 min at 250 F. The laminate produced will have excellent bond strength and water resistance.

Note:

The recommended minimum dry polymer pickup for this assembly is 65–70%.

Preliminary seal coating with a continuous film layer on the backside of the flannel serves as a barrier that prevents polymer strike through to the front side of the flannel when the polyurethane foam and the flannel are laminated.

Foam Splicing Spray Adhesive

(Rubber/Resin)

Kraton® 1107	50.0
Kraton® 1101	50.0
Sta-Tac®-R	100.0
1,1,1-Trichloroethane	360.0 *
Methylene Chloride	840.0 *
Sudan Red	1.0

---

* or to suit solids and visc. requirements

Properties:

Solids	25%
Visc. @ 72 F	800 cps

Note:

Above system is suited for foam-to-foam and scrim-to-foam bonding. Exhibits a soft glue line.

Polystyrene Foam Adhesives

	Formula No. 1	No. 2	No. 3	No. 4
	(Styrene Butadiene/Resin)			
Solprene 416-P	75	75	75	75
Ameripol 4503-C	25	25	25	25
Picco® 6100	75	75	75	75
Escorez® 2101	50	50	50	50
Champion Clay	200	200	200	200
Stanwhite 325	400	400	400	400

	No. 1	No. 2	No. 3	No. 4
		(cont'd.)		
Agerite D	2	2	2	2
Heptane	103	140	143	140
Rubber Solvent	103	—	—	46
Diisobutyl Ketone	—	140	—	—
Methyl Cyclohexane	—	—	61	—

	No. 5	No. 6	No. 7
	(Styrene Butadiene/Resin)		
Solprene 416-P	75	75	75
Ameripol 4503-C	25	25	25
Picco® 6100	75	75	75
Escorez® 2101	50	50	50
Champion Clay	200	200	200
Stanwhite 325	400	400	600
Agerite D	2	2	2
Shellflex® 371	—	20	30
Heptane	118	104	124
Isopentane	118	104	124

No. 8

(Styrene Butadiene/Resin)

Solprene 300	100
Dymerex	75
Antioxidant 2246	2
Hexane	660

ADHESIVES FOR LEATHER

Leather Adhesive

Formula No. 1

(Latex/Resin)

Natural Rubber Latex (60%)	100.0
Sodium EDTA (20% sol'n.)	2.5
Antioxidant	1.0
Methyl Cellulose (5% sol'n.)	2.0
Tackifying Resin Dispersion	10–20.0

No. 2

(Latex)

Natural Latex (60%)	167.0
Ethylene Diamine Tetraacetic Acid (20% sol'n.)	2.5
Antioxidant (50% dispersion*)	1.0

---

*   Poly-2,2,4-trimethyl-1,2-dihyroquinoline, e.g., **Flectol H.**

No. 3

(Rubber/Resin)

Kraton® 1101	100.0
Picco® LTP-135	37.5
Piccotex® 120	37.5
Antioxidant 330	0.5
DLTDP	0.5

Properties: Open time 60 min (no activation)

	Time after bond assembly		
180° Peel Strength	1 day	1 week	1 month
Canvas/Canvas*, pli	64	100	105
Canvas/PE**, pli	6	6	4
Canvas/Leather**, pli	12	10	10
Canvas/Rubber**, pli	12	12	16
Canvas/Wood**, pli	6	8	9
Canvas/Steel**, pli	11	11	12
Canvas/Glass**, pli	9	10	10

---

*   Instron jaw separation speed 2.0 in./min

** Instron jaw separation speed 0.5 in./min

Note:

The compound may be softened to facilitate contact or mating by adding 10–20 parts of a paraffinic oil. Some retained solvent is necessary on mating to insure formation of a bond across the glue line. For this reason mating open time is about 2 h. Heat treating or a solvent wipe may be used to activate the surface beyond the 2 h limit.

Flock-to-Leather Adhesives

	Formula No. 1	No. 2	No. 3	No. 4
		(Latex)		
Hycar® 2671	100.0	—	—	—
Geon® 460X1	—	100.0	—	—
Geon® 460X2	—	—	100.0	—
Hycar® 1572X45	—	—	—	100.00
Hydroxyethyl Cellulose	0.5	0.5	0.5	—
Good-Rite® K-718	—	—	—	0.25
Zinc Oxide Dispersion	—	—	—	3.00

Properties:

Brookfield Visc. RVF @ 20 rpm	3600	5000	3800	4900

No. 5

(Latex/Resin)

Hycar® 2671 (52%)	100.0
Carbopol® 934	0.5

Properties:

pH*	8.5

---

* Adjusted with 10% ammonium hydroxide before thickening with **Carbopol® 934**.

Leather-to-Plastic, Hot-Melt Adhesive

(Ethylene Vinyl Acetate)

Elvax 150	28.0
Elvax 240	18.0
Super Sta-Tac® 100	34.0
Polyethylene AC-8	5.6
Paraffin Wax (m.p. 150 F)	14.0
Antioxidant	0.4

Properties:

Ball & Ring Softening Pt.	100 C
Visc. @ 350 F	15,000 cps

Note:

The use of a primer, which is applied to the polyethylene or polypropylene substrates, will help improve the adhesion and strength of the bonded assembly.

Vinyl-to-Leather Adhesive

Formula No. 1

(Latex)

Geon® 450X20 (55%)	100.00
Methocel® HG-DGS (15,000 cP @ 5%)	1.25
Tetrasodium Pyrophosphate	0.50
Nopco® KOY (30% emulsion)	8.50

Procedure:

Coat the unfinished leather with the compounded latex blend using a #36 wire-wound rod.

Allow the coated leather to dry at room temperature. Laminate a 1-mil vinyl sheet to each piece of adhesive-coated leather at 136 psi and 180 F for 3 s.

No. 2

(Latex/Resin)

Chemigum N-S	100
SP-12	50
Monochlorobenzene	450
MEK	150

LAMINATING ADHESIVES

Laminating Adhesive

Formula No. 1

(Acrylic Resin)

Rhoplex® E-32	87.0
Nopco® DF-160 L (diluted 1:1 with water)	0.4
Acrysol® ASE-60 ⎫ Premix	3.0
Acrysol® ASE-95 ⎭	0.9
Water	3.9
RK-8	4.3
Ammonium Hydroxide (28–30% NH_3)	0.5
	(to pH 8.0)

No. 2

(Acrylic Resin)

Rhoplex® E-269	94.5
Nopco® DF-160 L (diluted 1:1 with water)	0.3
RK-8	4.7
Ammonium Hydroxide (28–30% NH_3)	0.5
	(to pH 7.5–8.0)

	No. 3	No. 4	No. 5
	(Acrylic Urethane)		
Purelast 169	55	55	55
POEA	40–50	—	—
THFUA	—	40–50	—
BZA	—	—	40–50
Photoinitiator	5	5	5

Uses:
  Nylon 6—Formula 3, 4, or 5
  Rigid Vinyl—Formula 4 or 5
  Cellophane—Formula 3
  Aluminum Foil—Formula 5

No. 6

(Resin/Rosin)

Ethyl Cellulose	10.0
Lewisol® 28	5.0
Dibutyl Phthalate	1.5
Ethanol	16.7
Toluene	66.8

No. 7

(Latex)

	Dry	Wet
Dow Latex 283 (45%)	70.00	155.6
Copolymer Tackifier Emulsion* (55%)	30.00	54.5
Polyacrylate Thickener	0.25	2.2

* Alkylated pentadiene/acetate resin.

No. 8

(Latex)

	Dry	Wet
Dow XD-8609.01	100.0	208.0
Magnesium Hydroxide	5.0	5.0
Elvanol 71-30	1.5	7.5

No. 9

(Latex)

	Dry	Wet
Dow XD-8260.04	100.0	200.0
Antimony Trioxide	5.0	5.0
Alumina Trihydrate	100.0	100.0
Magnesium Hydroxide	5.0	5.0

67.74% solids

Uses:
    Ignition deterrent.

No. 10

(Polypropylene/Resin)

Zonarez 7100	25.0
Microcrystalline Wax (m.p. 165 F)	50.0
Atactic Polypropylene	25.0

Properties:

Sealing Temp.	350 F
Visc. @ 350 F	71.5 cps
MVTR	0.06 g H_2O/100 in.^{2}–24 h
Adhesion:	
Kraft to Kraft	15 oz/in.
Kraft to Aluminum Foil	11 oz/in.
Kraft to Polyethylene	12 oz/in.
Kraft to Polypropylene	18 oz/in.

Uses:

A low-cost laminating adhesive for paper, film, and foil. This formulation has exhibited good barrier properties and creep resistance making it especially useful for packaging applications.

Vinyl Plastisol Sheet-to-SBR Sponge Laminating Adhesive

(Latex/Resin)

	Wet
Geon® 460X1	50.0
Isophorone	5.0
Hycar® 1562X103	60.0
Picconol® A600E	91.0

Procedure:

Apply the adhesive to both substrates, dry at 100 C (212 F) until all whiteness is removed, then bond the construction. After lamination, nip the bonded structure through squeeze rolls at 14 kPa (2 psi).

Flameproof Plastisol Laminating Adhesive

	Formula No. 1	No. 2
	(Phenolic Resin)	
Opalon 410	100	100
Santicizer 67	60	—
Santicizer 148	40	50
Santicizer 140	—	50

Note:

For maximum adhesion, stabilizers, fillers, etc., should not be used.

Flame-Retardant Plastisol Laminating Adhesives

	Formula No. 1	No. 2	No. 3
	(Vinyl Chloride)		
Geon® 121	100	100	100
Santicizer 148	—	55	110
Santicizer 711	110	55	—
Drapex 10.4	5	5	5
Mark 462	2	2	2
Magcarb L	20	20	20
Thermogard S	—	4	4

	No. 4	No. 5	No. 6
	(Vinyl Chloride)		
Geon® 128	100	100	100
Santicizer 148	—	80	80
Santicizer 711	80	—	—
Magcarb L	—	—	20

Thermogard S	—	—	4
Drapex 10.4	5	5	5
Mark 462	2	2	2

	No. 7	No. 8
	(Vinyl Chloride)	
Geon® 121	100	100
Santicizer 154	70	—
Santicizer 711	—	70
Stabilizers	7	7
Magcarb L	15	15
Thermogard S	5	5

Water-Resistant Laminating Adhesive

(Dextrin)

Water	51.77
Stadex 27	41.20
Salt	0.42
Soap	0.21
Urea-Formaldehyde Resin (65% solids)	6.20
Preservative	0.10
Defoamer	0.10

Procedure:

Cook @ 190–195 F for 20 min. Cool to 120 F. Add:

Alum Sol'n.   as required to adjust pH

Cool and dilute.

Properties:
Solids	42–45%
pH	5.5–6.0

Water-Soluble Laminating Adhesive

(Dextrin)

Water	84.0
Stadex 27	31.7
Borax (5 mole)	4.1
Defoamer	0.1
Preservative	0.1

Procedure:
Cook @ 190–195 F for 15 min. Cool and dilute if required.

Properties:
Solids   33–35%

Low-Curl Laminating Adhesive

(Dextrin)

Water	44.8
Clay (**ASP®-400** or equivalent)	5.0
Preservative	0.1
Defoamer	0.1
Sodium Nitrate	6.0
Borax (5 mole)	4.0
Stadex 92	40.0

Procedure:
Mix @ room temp. Cook 15 min @ 190–195 F. Cool and dilute if required.

Layflat Laminating Adhesive

Formula No. 1

(Dextrin)

1.	Water	36.20
2.	Defoamer	0.05
3.	Preservative	0.05

4. Sodium Nitrate	7.35
5. Ethylene Glycol	14.70
6. **Koldex 60**	17.15
7. **Stayco M**	9.80
8. Borax (10 mole)	2.45
9. Urea	12.25

Procedure:

Add #2–8 in the order listed, to the water @ room temp. Heat to 190 F. Hold at 190–195 F for 15 min. Cool below 140 F. Add #9. Dilute, if required, to adjust visc. Mix and cool.

Properties:

Refractometer Solids	52–53%
pH	$\approx$ 7–8
Brookfield Visc. @ 72 F–fresh	2500–3000 cps

No. 2

(Dextrin)

Water	31
Stadex 45B	38
Star-Dri 42R	13
Urea	18

Procedure:

Mix @ room temp. Heat to 190 F. Hold 15 min. Cool below 120 F.

Properties:

Refractometer Solids	63–64%
Brookfield Visc. @ 75 F	$\approx$ 3500 cps
pH	$\approx$ 9

Water-Resistant Tube-Winding and Laminating Adhesive

(Starch)

Water	70.0
Pearl Starch	3.9
Polyvinyl Alcohol (fully hydrolyzed grade)	8.1
Clay	17.9
Preservative	0.1

Procedure:
Cook 20 min @ 195–200 F. Cool below 120 F.

Kraft Laminating Adhesive

(Resin)

APP	60 (40–70)
Petroleum Resin	30 (25–40)
Microcrystalline Wax	10 (5–35)
BHT	0.2 (0–0.5)

BINDERS

Binder

(Styrene-Butadiene)

Kraton® 1101	100.0
Whiting	700.0
Paraffinic Oil	150.0
Antioxidant 330	0.3
DLTDP	0.3

Properties:

Melt Flow, cond. E	15 g/10 min
Hardness (Shore A)	55
Specific Gravity	1.8

Note:

When applying the compound by calendering, 1–2 phr stearic acid may be included to prevent sticking. The compound may also be extrusion coated.

Uses:

Integral backing for carpet tiles, sound insulation.

Binder for Ground Cork

Formula No. 1

(Gelatin/Acrylic)

Ground Cork	100
Glycerin	5
Gelatin (≈ 500 Bloom g)	10
Water	15
Acronal 500 D	5–10
Hexamethylene Tetramine (15% sol'n.)	2

Procedure:

Mix the ground cork with the gelatin solution containing glycerin. Then incorporate **Acronal 500 D** and apply the hardener solution by spraying. Dry the made-up mixture to a moisture content of approximately 8%. The mixture is processed in extruders at approximately 120 C. The gelatin cures at this temperature. After they have been sanded the sections are cut into discs.

Note:

Predrying is particularly important for obtaining satisfactory results in processing the mixture in extruders. Moreover, it is necessary to add paraffin wax or mineral oil. The amount to be added depends on the requirements imposed. Normally 5% calculated on the amount of ground cork is sufficient.

Gelatin-bonded moldings are not resistant to water and oil. Glycerin is often used for plasticizing. The gelatin can be hardened by formaldehyde solution or substances which give off formaldehyde (e.g., hexamethylene tetramine). The flexibility of the pressed cork decreases as a result of this treatment. This cannot be prevented even by adding a larger amount of glycerin. However, a softer material can be obtained by adding a polymer dispersion.

Uses:

The bonded materials are produced in the form of sheets, panels, or round rods, which are processed to wedges for shoe soles, shoe inserts, gaskets, and similar articles.

No. 2

(Glycerin/Acrylic)

Ground Cork	100.0
Hide Gelatin ($\approx$ 500 Bloom g)	3.4
Acronal 500 D	2.7
Glycerin	6.0
Water	14.6
Formaldehyde (15% sol'n.)	0.8

Procedure:

The formaldehyde solution is sprayed onto the ground cork/binder mixture shortly before pressing.

Uses:

For pressed cork discs for pilferproof closures.

No. 3

(Glycerin/Acrylic)

Ground Cork	100.0
Hide Gelatin (≈ 500 Bloom g)	3.4
Acronal 500 D	6.0
Glycerin	8.7
Water	10.9
Formaldehyde (15% sol'n.)	0.8

Note:

In order to achieve uniform distribution on the ground cork, the dispersion must be diluted with water according to the quantity of binder required. After the binder has been applied, the compound is carefully dried at 70–80 C to a total water content of 6–8%. Pressed cork is produced in heatable molds.

When non-crosslinkable dispersions are used, the mold must be allowed to cool after pressing so that the molding retains its shape. In using thermally crosslinkable dispersions, care must be taken that the crosslinking temperature is also attained in the core of the molding.

Uses:

For pressed cork discs for crown caps.

No. 4

(Polyvinyl Propionate)

Ground Cork	100
Propiofan 5 D	20
Water	2

No. 5

(Acrylic)

Ground Cork	100
Acronal 330 D	40

Note:

The crosslinking schedule for the dry film of **Acronal 330 D**, i.e. 3 min at 150 C, may serve as a guide for crosslinking.

The ground cork/resin mixture should be carefully predried to a total moisture content of approximately 8% in order to prevent the formation of steam blisters during pressing. Steam blisters are particularly liable to cause trouble in thick moldings. The temperature curing predrying should be controlled in such a manner that the condensation resin does not dry prematurely. The hardener solution is sprayed onto the predried ground cork/binder mixture shortly before pressing.

Normally, lockable molds are used, and curing is effected in a heating chamber. The heating time must be determined by trials in each individual case in order to achieve uniform strength in the molding.

No. 6

(Urea Resin)

Ground Cork	100.0
Urecoll 181	15.0
Ammonium Chloride (15% sol'n.)	1.5

Properties:
The pot life of the binder/hardener mixture is:
$\approx$ 27 min @ 20 C
$\approx$ 10 min @ 30 C
$\approx$ 16 s @ 100 C

No. 7

(Urea Resin)

Ground Cork	100.0
Urecoll 118	15.0
Ammonium Chloride (15% sol'n.)	1.5

Properties:
The pot life of the binder/hardener mixture is:
$\approx$ 10.5 h @ 20 C
$\approx$ 3.5 h @ 30 C
$\approx$ 35 s @ 100 C

No. 8

(Urea Resin)

Ground Cork	100.0
Urecoll 270	20.0
Water	6.5
Ammonium Chloride (15% sol'n.)	3.0

Properties:
The pot life of the binder/hardener mixture is:
$\approx$ 17 min @ 20 C
$\approx$ 7 min @ 30 C
$\approx$ 15 s @ 100 C

Note:
Like the liquid **Urecoll** types, the **Urecoll** powder types can be made to react at different rates. **Urecoll 270** in Formula No. 8 reacts faster than **Urecoll 237** in Formula No. 9

No. 9

(Urea Resin)

Ground Cork	100.0
Urecoll 237	20.0
Water	6.5
Ammonium Chloride (15% sol'n.)	3.0

Properties:
The pot life of the binder/hardener mixture is:
$\approx$ 8.5 h @ 20 C
$\approx$ 4 h @ 30 C
$\approx$ 38 s @ 100 C

No. 10

(Urea Resin)

Ground Cork	100
Urecoll 237	20
Casein Sol'n.*	3
Ammonium Chloride (15% sol'n.)	3

---

* Casein Sol'n.:

Casein	16
Water	80
Ammonia (25% sol'n.)	5

No. 11

(Urea Resin)

Ground Cork	100.0
Urecoll 118	7.5
Wheat Flour	1.0
Water	7.0
Diethylene Glycol	3.0
Ammonium Chloride (15% sol'n.)	0.7

Note:

Urea resins can be modified with casein (Formula No. 10) or with wheat flour (Formula No. 11). A certain degree of elasticity is thus obtained.

Solid Rocket Propellants Binder

HC Polymer 434	84.20
Mapo	2.22
Erla-0500	1.78
Asbestos Floats	10.30
Thixcin E	1.00
Iron **Octasol**	0.50

Procedure:

By making the binder thixotropic, **Thixcin E** allows the liner to be applied in one coat. The liner is precured 10 h @ 135 F (57 C) before the propellant is cast against it, and is then fully cured, along with the propellant for another 140 h.

Note:

The liner in a rocket motor serves three purposes: (1) bonds the propellant to the case material, (2) protects the case from the hot gases during tailoff, and (3) protects the propellant (adjoining the case) from aerodynamic heating.

Leather Scrap Binder

	Formula No. 1	No. 2
	——— (Latex) ———	
Leather Fiber	100	100
Hycar® 1562X103 (10% T.S.)	30	60
Nalco 600	8	12
Water for Process	5000	5000

Nonwoven Textile Binder

	Formula No. 1		No. 2	
	——————— (Latex) ———————			
	Dry	Wet	Dry	Wet
Hycar® 2600X120	60.0	119.5	—	—
Hycar® 2679	—	—	60.0	120.7
Water	—	272.5	—	271.3
Oxalic Acid	0.7	8.0	0.7	8.0

After saturation, nip at 20 lb of pressure and immediately dry for 5 min at 135 C (275 F).

	No. 3		No. 4		No. 5	
	(Latex)					
	Dry	Wet	Dry	Wet	Dry	Wet
Hycar® 2600X104	60.0	116.7	60.0	116.7	60.0	116.7
Permafresh LF	—	—	3.0	3.0	6.0	6.0
Ammonium Chloride	—	—	0.3	3.0	0.6	6.0
Water	—	275.3	—	290.8	—	306.3
Oxalic Acid	0.7	8.0	0.7	8.5	0.8	9.0

After saturation, nip at 20 lb of pressure and immediately dry for 5 min at 135 C (275 F).

No. 6

(Water-Based, Acrylic)

Ethyl Acrylate	400.00
Methyl Methacrylate	100.00
N-Methylol Acrylamide (48% sol'n.)	10.40
Methacrylic Acid	5.00
Sodium Bisulfite (3% sol'n.)	40.00
Sodium Persulfate	1.25
Siponic® F-400	43.00
Water	370.00

Procedure:

Charge 220 ml water and sodium persulfate to a one-liter resin kettle reactor. Heat reactor contents to 60 C while purging with nitrogen. Prepare a solution of surfactant and N-methylol acrylamide in 150 ml of water. Form preemulsion by gradually adding monomer blend to surfactant solution while stirring vigorously. Transfer preemulsion to an addition funnel. When temperature of reactor reaches 60 C, add 5.0 ml of sodium bisulfite solution and begin addition of preemulsion. Schedule preemulsion addition for 3–4 h. Maintain temperature between 60–65 C. Add sodium bisulfite solution incrementally throughout the reaction. When all ingredients are added, continue heating reactor for 30 min. Cool and discharge. Adjust latex to pH 7 with ammonia.

Latex Properties:

Solids	55.5%
pH	7.0
Visc. (Brookfield #1 spindle/6 rpm)	350 cP

No. 7

(Water-Based, Latex)

Ucar® Latex 874	60.00
Water	30.00
Cellosize® QP (4% aq. sol'n.)	6.00
Aerosol® 18	3.20
Sipex® UB	0.80

No. 8

(Natural Rubber)

Natural Rubber Latex (60%)	167.0
Texofor FN36 (20%)	2.0
Sulfur Dispersion (50%)	4.0
Zinc Diethyl Dithiocarbamate Dispersion (50%)	3.0
Zinc 2-Mercapto Benzothiazole Dispersion (50%)	1.5
Antioxidant Dispersion (50%)	2.0
Zinc Oxide Dispersion (50%)	6.0
Soft or Distilled Water to adjust total solids	as required

Suggested cure 3–8 min at 120 C in hot air.

No. 9

(Delamination-Resistant, Natural Rubber)

Natural Rubber Latex (LA type, 60%)	167.0
Nonionic Stabilizer Sol'n. (20%)	1.3
Formaldehyde Sol'n. (40%)	2.2
Sulfur Dispersion (50%)	4.0

Zinc Oxide Dispersion (50%)	6.0
Zinc Diethyl Dithiocarbamate Dispersion (50%)	3.0
Zinc 2-Mercapto Benzothiazole Dispersion (50%)	1.5
Antioxidant Dispersion (50%)	2.0
Polyvinyl Methyl Ether Sol'n. (10%)	20.0
Water as necessary	(e.g., 20% DRC)

Properties:

Coagulation Temperature	≈ 32 C
Minimum Shelf Life @ 20 C	1 week

Foundry Hot-Box Binder

(Resin)

Lake Sand*	2000.0
Water	2.8
Furset Catalyst	1.6
Furset Modifier	8.8
Furset 138	40.0
Core Oil**	2.0

---

*   88–90% through 40 mesh and retained on 70 mesh screens.
** Optional

Procedure:
Slurry water and **Furset Catalyst**. Add to sand and mix one minute.
Add **Furset Modifier**. Mix one minute. Add **Furset 138**. Mix 2
min. Add oil. Mix one minute and discharge.

SEALANTS

Clear, Weatherable Sealants
Formula No. 1

(Rubber/Resin)

Kraton® G-1652	67.0
Kraton® GX-1701	33.0
Arkon® P-85	167.0
Kristalex® 1120	67.0
Silane A-189	2.3
Irganox® 1010	1.7
Tinuvin® 327	0.7
Tinuvin® 770	0.7
Tolusol 25	109.0
Propyl Acetate	37.0

6Properties:*

Hardness, Shore A	26
180° Peel Strength:	
against glass, original**	30 pli
against glass, after 1 week soaking in water**	25 pli
against glass, after 200 h exposure in xenon arc weatherometer**	13 pli
Color change or surface cracks after 1000 h in xenon arc weatherometer	Slight yellowing
Durability cycling with 50% joint movement**, —15–70 C	Passes against glass and aluminum
Tack after 72 h drying time**	Passes

---

* Measured after one month drying time
** Measured according to TT-S-00230C.

No. 2

(Solvent-Based/Rubber/Resin)

Kraton® G-1652	67.0
Kraton® GX-1701	33.0
Arkon® P-85	167.0

Kristalex® 1120	67.0
Silane A-189	2.3
Antioxidant	1.7
UV Inhibitors	1.4
Toluene/Hexane	109.0
Propyl Acetate	37.0

Solvent-Based Clear Sealants

Formula No. 1

(Rubber/Resin)

Kraton® GX-1657	100.0
Arkon® P90	147.0
Kristalex® 3085	67.0
Irganox® 1010	1.7
Tinuvin® P	0.8
Tinuvin® 770	0.8
Silane® U-189	2.7
Cab-O-Sil®	13.0
Toluene	111.0

Properties*:
Hardness, Shore A durometer
Instantaneous	35
10 s	25

180° Peel on Glass
Peel strength	30 pli
	5250 N/m
Failure mechanism	Cohesive

---

* Properties of fully cured sealant, measured according to Federal Spec. TTS-00230C.

Precautions:

Kraton® thermoplastic rubber crumb can accumulate electrostatic charges when rubbed, chafed, or abraded. Equipment should provide a means for dissipating any charges that may develop. Compounding of **Kraton®** rubber crumb in high-shear equipment can cause the temperature to rise. **Do not allow the temperature to exceed 450 F. Maintain a fire watch if 450 F is reached.**

	No. 2	No. 3	No. 4	No. 5
	(Rubber/Resin)			
Kraton® GX-1657	100	100	100	100
Arkon® P-85	167	167	167	167
Kristalex® 3085	67	—	—	—
Kristalex® 3100	—	67	—	—
Kristalex® 1120	—	—	67	—
Kristalex® 5140	—	—	—	67
Toluene	120	120	120	120
Properties:				
Clarity	Clear	Clear	Opaque	Opaque
Slump after 1 wk at 70 C	3	2.5	—	—

	No. 6	No. 7	No. 8	No. 9
	(Rubber/Resin)			
Kraton® G-1652	67	67	67	67
Kraton® GX-1701	33	33	33	33
Arkon® P-85	167	167	167	167
Kristalex® 3085	67	—	—	—
Kristalex® 3100	—	67	—	—
Kristalex® 1120	—	—	67	—
Kristalex® 5140	—	—	—	67
Toluene	120	120	120	120
Properties:				
Clarity	Clear	Clear	Clear	Clear
Slump after 1 wk at 70 C	1.5	1	0	0

No. 10

(Rubber/Resin)

Super Nirez™ 5100	200.00
Kraton® G-1652	100.00
Indopol® H-300	20.00
Irganox® 1010	0.50

Tinuvin® 328	0.25
Tinuvin® 770	0.25
Hexane	99.50
Toluene	30.00
Propyl Acetate	43.50

Properties:

Solids	65%

Note:

This system exhibits good clarity and resistance to ultraviolet light discoloration.

Adhesion to glass can be improved by the addition of a coupling agent such as a functionalized silane.

No. 11

(Rubber/Resin)

SEBS Block Polymer	100.0
Super Nirez™ 5100 or 5120	200.0
Irganox® 1010	0.5
Hexane	92.0
Toluene	28.0
Propyl Acetate	40.0

Properties:

Solids	65.3%

No. 12

(Polybutene/Butyl Rubber)

Butyl 268	5.7
Vistanex L-100	10.6
Super Sta-Tac® 100	8.2
Indopol® H-100	16.3
Atomite®	8.2
IT-X	36.0
Mineral Spirits	14.9

Properties:

Solids   85.0%

Procedure:

Charge butyl rubber/polyisobutylene to a sigma-blade mixer (or suitable mixing equipment) with solvent. Mix. Add resin, $^1/_2$ polybutene, talc, $^1/_2$ calcium carboante. Mix. Add remaining calcium carbonate. Mix. Add remaining polybutene. Mix. Mix until system is homogeneous.

No. 13

(Polybutene/Butyl Rubber)

AD-50	200
Indopol® H-100	125
Atomite®	500
Asbestine 3X	125
Marble Dust	125
Titanox® 2101 or Kronos RN 57	12
Super Sta-Tac® 80	20
Bentone® 34 or 500 or Thixcin GR or	
Cab-O-Sil® M5/Aerosil® 200	10 or 25
Mineral Spirits (odorless)	50

No. 14

(Ethylene Vinyl Acetate/Nitrocellulose)

RS Nitrocellulose (30–35 cP)	50.0
Vynathene EY 902-30	50.0
Isopropyl Alcohol	21.5
n-Butyl Alcohol	158.5
n-Butyl Acetate	307.2
n-Propyl Acetate	35.7
Isopropyl Acetate	17.1
Toluene	360.0

Procedure:

The recommended procedure for preparing a **Vynathene**/nitrocellulose lacquer is to dissolve the VAE pellets in the solvent first, then add the appropriate amount of nitrocellulose solution.   Preparation will be considerably lengthened, if the **Vynathene** is added last.

Properties: (Film properties of a **Vynathene VA**-modified nitrocellulose lacquer applied to 5-ply maple panels.   One thin coat of lacquer was applied as a sealer and dried for 30 min.   It was scuff sanded and followed with two full coats with a 30-min drying between coats.

Nonvolatile Content	10%
Visc. #4 Ford Cup @ 75 F	17 s
Print Resistance[1]	No print
Impact Resistance[2], in.-lb	Pass 50
	Fail 60
Scratch Resistance[3], g	Pass 500
	Fail 600
Cold Check Resistance[4], g	No crack
Sward Hardness	38

---

[1]   4 psi after 16 h drying of topcoat
[2]   Gardner Tester Model IG-1120
[3]   Hoffman Tester
[4]   25 cycles

Note:
    A hard maleic modified rosin such as **Pentrex 28** is sometimes added to nitrocellulose finishes to reduce the cost and increase the nonvolatile content at the application viscosity.

High-Temperature Structural
Plastisol Sealant

(Vinyl Resin)

Bakelite QYOH-2	100.0
Duramite	100.0
Flexol® 10-10	40.0
Flexol® 10-A	25.0
Super Multiflex	15.0
Dyphos	2.0
Silane A-1120	0.5

High-Temperature Fusing Plastisol Sealant

(Polyvinyl Chloride/Resin)

PVC Homopolymer Dispersion Resin	100
Calcium Carbonate Filler	125
Pyrogenic or Precipitated Silica	7
Kodaflex PA-5	50
Kodaflex DOTP	60
Epoxidized Soybean Oil	5
Dibutyl Tin Maleate or Dibutyl Tin Laurate	1
Dibasic Lead Phthalate	3
Methyl Alcohol or Ethylene Glycol	1

Properties:
Fusing Temp.   177 C (350 F)

Self-Adhering Sealant Compound

(Silicone)

RTV-31U	47.3
Burgess #30 Clay	30.4
Thixcin R	5.2
Silane A-1110	1.5
Mineral Spirits (odorless)	15.1

Low-Temperature Plastisol Sealant

(Vinyl Resin)

Bakelite VLFV	65
Bakelite QYPM	35
Bakelite VMCC	20
Flexol® 10-10	30
Flexol® 380	40
Flexol® A-26	55
EX-1 (aliphatic hydrocarbon)	25
Dyphos	2

Bakelite ERL-4289	5
Silane A-1112	1
Camel-Kote or **Duramite**	180
Ircogel 900	10

Low-Temperature Fusing Sealant

(Polyvinyl Chloride Resin)

PVC/PVA Copolymer Dispersion Resin	100
Calcium Carbonate	100
Talc	25
Pyrogenic or Precipitated Silica	6
Kodaflex PA-5	50
Kodaflex HS-3	60
Epoxidized Soybean Oil	5
Dibutyl Tin Maleate or Dibutyl Tin Laurate	1
Dibasic Lead Phthalate Stabilizer	3
X-970	4
Methyl Alcohol or Ethylene Glycol	1

Properties:

Fusing Temp.   135–149 C (275–300 F)

Note:

This formulation has demonstrated vinyl bonds stronger than the cohesive strength of the vinyl on cold-rolled steel and **Bonderite 40** phosphatized steel, under fusion conditions of 130 C (265 F) for 30 min or 149 C (300 F) for 20 min.

Rubberized Asphalt Sealants

	Formula No. 1	No. 2	No. 3	No. 4
	(Oil-Based, Styrene Butadiene)			
Solprene 1205	100	100	—	50
Solprene 406	—	—	100	50
Highly Aromatic Oil	100	50	100	100
Airblown Asphalt	467	1800	900	467
Polygard	2	2	2	2

	No. 1	No. 2	No. 3	No. 4
		(cont'd.)		
Carbowax 6000	5	5	—	—
Zinc Oxide	5	5	—	—
Stearic Acid	3	3	—	—
Bismate	2	2	—	—
MBTS	2	2	—	—
Sulfur	1	1	—	—

Procedure:

Mix cycle in Baker-Perkins mixer: Masterbatch rubber, $1/3$ of the oil, and all other ingredients except asphalt. Mix at 180 F. After 30 min, add remaining oil and asphalt. Mix until thoroughly dispersed then raise temperature to 280 F for 30 min to effect cure.

Aircraft Sealants

Formula	No. 1	No. 2	No. 3	No. 4	No. 5
LP-32 Polymer	100.0	100.0	100.0	100.0	100.0
Titanox® RA-50	—	50.0	—	—	—
EH-330	—	—	1.2	—	—
Maleic Anhydride (25% in cyclohexanone)	—	—	2.0	—	—
Durez 10694	5.0	—	—	5.0	—
SRF #3	30.0	—	—	40.0	30.0
Magnesium Oxide	—	—	4.0	—	—
Sulfur	0.1	—	—	0.1	—
Stearic Acid	1.0	—	—	1.0	—
Cumene Hydroperoxide	—	—	6.0	—	—
Tellurium Oxide (50% in dibutyl phthalate)	—	4.0	—	—	—
Lead Oxide (50% in **Aroclor 1254**	15.0	—	—	—	—
Ammonium Chromate (43% in water)	—	—	—	15.0	—
Cab-O-Sil® M-5	—	—	20.0	—	—
Sodium Stearate	—	5.0	—	—	—
Manganese Oxide-D Grade	—	—	—	—	3.0

Polymercaptan Sealant

Formula No. 1

(Resin)

Dion DPM-1002	42.0
Chlorowax 500C	4.3
Dioctyl Phthalate	6.3
Silane Y-4523	0.3
Thixcin R	3.6
Titanium Dioxide (DR-60)	6.3
Calcium Carbonate	32.0
Molecular Sieve (4A)	0.4
Calcium Peroxide	4.2
Barium Oxide	0.6

Procedure: One-step formulation (constant low-humidity conditions)

Combine **DPM-1002** resin, plasticizer, and **Thixcin R.** Blend with a high-shear mixer until a temperature of 130–140 F (54–60 C) is reached. Add silane and blend in with slow-speed mixer. Add remaining ingredients and blend the entire formulation with a slow-speed mixer. Pass over a tight three-roll mill to a Hegman value of six or less. Package in moisture-impermeable containers.

No. 2

(Resin)

Base Component:	
Dion DPM-1002	100.00
Dioctyl Phthalate	35.00
Emersol® 132	1.80
Thixcin E	15.00
Super Multiflex	110.00
Titanox® RA-NC	10.00
Thermax MT	1.00
Activator:	
Lead Peroxide (med. cure)	7.50
Dioctyl Phthalate	6.75
Emersol® 132	0.75

Nondrying Sealant

(Polybutene)

Amoco® Polybutene H-300	31.75
Soya Fatty Acid	0.54
Short Fiber Asbestos	31.75
Calcium Carbonate	31.75
Diatomaceous Silica	4.21

Procedure:

In the laboratory the sealant is mixed in a sigma-blade mixer. Additions to the mixer are made in the order listed. The entire mass is then mixed for 1 h after the last addition.

Sealant for Potable Water Tanks

(Resin)

Thiokol LP-2	100
Titanox® RA-50	5
Calcene TM	25
Durez 10694	5
Stearic Acid	1
Catalyst*	15

* Catalyst Recipe:	
PhO_2 (med. grade)	50
Aroclor 1254	54
Stearic Acid	5

Aquarium Sealant

	Formula No. 1	No. 2
	(Butyl Rubber)	
Enjay Butyl LM 430	50	50
Omya® BLH	60	—
HiSil 215	—	20
Indopol® H-100	—	30
Silane A-187	4	—

GMF	3	—
Lead Oxide (VFC 0.33μ)	—	8
Toluene	12	19

Insulated Glass Sealant

Formula No. 1

(Butyl Rubber)

A	**Enjay Butyl LM 430**	75.0
	Toluene (dry)	20.0
	Silane A-187	4.0
	Omya® BLH	75.0
	GMF	3.5
	AgeRite White	1.5
	Aroclor 1254	7.5
B	**Enjay Butyl LM 430**	25.0
	Toluene (dry)	20.0
	Aroclor 1254	7.5

No. 2

(Butyl Rubber)

A	**Enjay Butyl LM 430**	60
	Toluene	15
	Thixseal 436	6.0

Thoroughly mix @ a min. temp. of 158 F (70 C).

Epoxy Silane	4.0
Indopol® H-1900	10.0
AgeRite White	1.5
p-Quinone Dioxime	3.5
Omya® BLH	75.0
Toluene	30.0

Stir into above, then paint mill for good dispersion.

B Enjay Butyl LM 430	40.0
Lead Oxide	7.5
Cab-O-Sil® M5	5–10.0
Epon® 872	0–10.0
Toluene	20–40.0

Note:

Epon® 872 can slow bonding rate so its contribution to flow control should be balanced against the bonding rate equipment.

Glass Window Sealant

	Formula No. 1	No. 2
	(Polybutene/Resin)	
Butyl Elastomer	15.6	16.6
HAF N330	23.5	25.5
Phenolic Resin	29.5	25.6
Amoco® Polybutene H-1900	23.5	17.5
Amoco® Resin 18-290	—	5.9
Ethylene Propylene Rubber	7.9	8.9

Membrane Capping Sealant

(Butyl Rubber/Resin)

Polysar Butyl XL-20	50
Polysar Butyl 402	50
FEF Black	120
HiSil EP	40
Resin ST 5115	20
Amberol ST 149	20
Indopol® H-300	100

Hand-Mixed Sealant Molding Compound

	Formula No. 1	No. 2
	(Butyl Rubber)	
Enjay Butyl LM 430	60.0	40.0
Calcene TM	20.0	—
Omya® BLH	100.0	—
Titanium Dioxide	5.0	—
Bayol 35	15.0	8.0
GMF	3.5	—
2-Pyrrolidone	2.0	—
Manganese Dioxide	—	10.0
Neodecanoic Acid	—	0.5

Work life ≈ 120 min, overnight cure

CONTACT CEMENT

Neoprene Cement

(Rubber/Resin)

A **Neoprene AD-20**	100.0
Zinc Oxide	5.0
Magnesium Oxide	4.0
Ionol	2.0
Toluene	100.0
Gasoline	233.0
B **SP-134**	45.0
Magnesium Oxide	4.5
Water	2.0
Toluene	154.5

Procedure:
Combine Parts A and B.

Sprayable Contact Adhesive

(Rubber/Resin)

Kraton® D 1101	100
Pentalyn® H	125
Picco® 6140	25
Cellolyn® 21	10
Antioxidant	1
Hexane	870
Toluene	370

General-Purpose Contact Cement

Formula No. 1

(Solvent-Based, Rubber/Resin)

Neoprene AD-20	100.0
Magnesium Oxide	4.0
Zinc Oxide	5.0

Antioxidant	1.0
Hexane	300.0
Methyl Ethyl Ketone	114.0
Toluene	57.0

Resin Prereaction:

Varcum® 875 or Super Beckacite® 1054	45.0
Magnesium Oxide	5.0
Water	0.5
Toluene	30.0
Hexane	15.0

Properties:

| Solids | 24% |
| Visc. | 1000–1500 cps |

Procedure:

Dissolve rubber in solvent with magnesium oxide, zinc oxide, and antioxidant. Prepare resin prereaction by dissolving resins in solvent. When totally dissolved, add magnesium oxide, then water. Mix for at least 2 h to insure maximum reaction with magnesium oxide. Mix rubber base with resin prereaction.

	No. 2	No. 3	No. 4
		(Rubber)	
Natural Rubber	100.0	100.0	100.0
Zinc Oxide	5.0	5.0	5.0
Sulfur	2.5	2.5	2.5
Altax	1.0	1.0	1.0
Diphenyl Guanidine	0.3	0.3	0.3
Antioxidant	2.0	2.0	2.0
ST-5115	100.0	—	—
ST-5135	—	100.0	—
SP-553	—	—	100.0
Hexane		to 20% solids	

No. 5

(Latex)

	Dry	*Wet*
Dow Latex 283 (45%)	70.0	155.6
Aliphatic Petroleum Tackifier Emulsion (45%)	30.0	66.2
Polyacrylate Thickener	0.5	4.5

	No. 6	No. 7
	(Paraffinic Petroleum/Resin)	
FEF	—	30
HAF	—	30
Flexon 875	—	5
Flexon 765	10.00	—
Stearic Acid	1.00	1
Amberol ST-137X	3.00	3
Antioxidant 2246	1.00	1
Diethylene Glycol	2.00	—
Maglite D	0.25	2
Zinc Oxide	—	3
Litharge	10.00	—
TMTDS	—	1
MBTS	—	2
NA-22	0.75	—

Properties:

Press cured as shown at 307 F, min	60	40
Hardness, Shore A	62	68
300% Modulus, psi	560	1350
Tensile Strength, psi	1960	1920
Elongation, %	720	500

	No. 8	No. 9	No. 10	No. 11
	(Styrene Butadiene/Resin)			
Solprene 411	100	100	100	100
Dymerex	100	40	—	—
Pentalyn® H	—	60	40	50
Picco 6100 $1^{1}/_{2}$	—	—	60	50
Textile Spirits	200	200	200	200
Toluene	200	200	200	200

	No. 12	No. 13
	(Styrene Butadiene/Resin)	
Solprene 406	100	100
Dymerex	100	60
Pentalyn® H	—	40

Properties:
  T-Peel Strength on 30 oz cotton duck

		No. 12	No. 13
lb/in.		56	61
kN/m		9.8	10.7
Upper Temp. Limit,			
100 g/in. static T-peel text,	F	240	230
	C	116	110

	No. 14	No. 15
	(Styrene Butadiene/Resin)	
Solprene 406	100	100
Zonarez 7100	50	—
Picco® 6110	50	50
Neville LX-685, 125	—	50
Toluene	400	400

	No. 16	No. 17
	(Styrene Butadiene/Resin)	
Solprene 411	100	100
Zonarez 7100	50	50
Neville LX-685, 125	—	50
Picco® 6110	50	—
Irganox® 1076	2	2
Toluene	400	400

EPOXY ADHESIVES

Epoxy Adhesive

Formula No. 1

Epon® 1007	25
Dicyandiamide	2
Dioxitol	76
Carbopol® 941	2

Procedure:

Dissolve the **Epon® 1007** in the **Dioxitol.** Disperse the **Carbopol®** resin slowly until homogeneous. Keep agitating and add the dicyandiamide. Stir for 1 h.

No. 2

Epoxy Resin	75.0
Epoxide #7	7.5
RG 144 Asbestos	3.5
325-Mesh Silica	14.0
Dion Hardener 38	35.0

Note:

This adhesive will adhere well to various substrates including steel, aluminum, wood, and concrete. A serviceable bond is produced after overnight curing at 77 F.

No. 3

(Polyamide)

Polyamide Systems	
A Bisphenol A Diglycidyl Ether	100
Additive	20
Asbestos Fibrils	5
Powdered Aluminum	50

B Polyamide-Amine	59
Tris (Dimethylaminomethyl) Phenol	3
Asbestos Fibrils	3

Cure @ 73 F for 7 days

AEA Systems:

A Bisphenol A Diglycidyl Ether	100.0
Additive	15.0
Asbestos Fibrils	5.0
Aluminum Powder	50.0

B AEA	19.3
Asbestos Fibrils	3.0

Cured @ 73 F overnight, postcured at 212 F for 2 h

DICY Systems—One Part

Bisphenol A Diglycidyl Ether	100.0
Additive	15.0
DICY Powder	10.0
Diuron	7.5
Asbestos Fibrils	5.0
Aluminum Powder	50.0

Cured @ 250 F for 30 min.

No. 4

Liquid Epoxy Resin	100
Diethylene Triamine	8
Weston™ EGTPP	25

Cure Schedule: 20 h @ room temp. plus 2 h @ 100 C.

No. 5

Liquid Epoxy Resin (DGEBA of EEW 190)	100
Jeffamine D-230	30
Accelerator 398	5
Aluminum Powder	30
Cab-O-Sil®	3

Procedure:

Components 1 and 2 should be mixed thoroughly. Pot life of this mixture is 6–10 h. When ready to apply to concrete surface, add Component 3. A gel-like material of putty-like consistency will instantly form. The material may then be troweled onto vertical or horizontal surfaces to the thickness desired.

	No. 6	No. 7	No. 8
A **LP-3**	100	125	—
LP-2	—	—	15
EH-330	20	—	—
DMP-10	—	—	10
Triethylene Tetramine	—	20	—
B **Epon® 828** (or equivalent)	200	—	—
Araldite 502	—	200	—
Epon® Adhesive VIII	—	—	24
Surfex MM	200	—	—
Mixing Ratio by wt. (based on 100 pbw Part A)	100/333	100/138	100/96

Properties:

	No. 6	No. 7	No. 8
Application	Brush, trowel roller coat	Brush or dip coat	Brush or trowel
Consistency	Paste	Very fluid	Viscous
Pot Life @ R.T.	30 min	30 min	

	No. 6	No. 7	No. 8
		(cont'd.)	
Cure	4–5 h @ R.T. for handling; 5–7 days for high strength; 15 min @ 220 F for accelerated cure.	5–7 days @ R.T. (or heat accelerated)	1.5–2.5 h @ 165–185 F

Uses:

(No. 6): General-purpose (low toxicity) for bonding aluminum, steel, tin, wood, glass, ceramics, some plastics also used as a pipe sealant.

(No. 7): Flexible, impact resistant exhibits good adhesion to steel, aluminum, wood, ceramics, and glass.

(No. 8): Bonding carbon to hard, anodized aluminum.

	No. 9	No. 10
A **LP-3**	100.0	100.0
EH-330	—	12.5
DMP-10	—	12.5
EC-1	38.4	—
Duramite	—	187.5
Asbestine	—	125.0
B **Epon® 828** (or equivalent)	120.0	250.0
Epon® 1004*	80.0	—
Allyl Glycidyl Ether	25.6	—
Duramite	—	375.0
Mixing Ratio by wt. (based on 100 pbw Part A)	100/163	100/143

---

* The solid epoxy is dissolved in the liquid epoxy at 300 F. Then the allyl glycidyl ether is incorporated into the blend while it is still fluid. The solution is cooled before application.

Properties:

Application	Brush	Trowel
Consistency	Fluid	Paste
Cure	5–7 days @ R.T. (or heat accelerated)	5–7 days R.T. (or heat accelerated)

Uses:

(No. 9): Brushable for bonding steel, aluminum, glass, ceramic tile.

(No. 10): Unglazed tile for bonding a variety of materials including tile to plaster and concrete surfaces both vertical and horizontal.

		No. 11	No. 12
A	**LP-3**	—	100
	LP-2	100	—
B	Epoxy	500	1000
	Resorcinol	100	—
	Hexamethylene Tetramine		
	(45% sol'n. in water*)	125	—
	Milled Glass Fibers ($^1/_{32}$ in.)	—	100
	Dimethyl Aminoethyl Alcohol*	—	40
	Triethylamine*	—	40
Mixing Ratio by wt. (based on			
100 pbw Part A)		100/750	100/1180

---

* The amine(s) must be packaged separately as a third component.

Properties:

Application	Brush	Brush
Consistency	Fluid	Fluid
Pot Life @ R.T.	1.5 h	2.5 h
Cure	8 days @ 80 F or	8 days @ 80 F
	1 h @ 200 F	or 1 h @ 200 F

Uses:

(No. 11): Elevated temperature service bonds steel, aluminum, ceramic and glass exhibiting good shear strength at temperatures as high as 250 F.

(No. 12): High impact resistance and flexibility, combined with high bond strength.

Thickening Epoxy Adhesives

Formula No. 1

Carbitol® or Cellosolve	76.0
Epon® 1007	25.0
Carbopol® 941	2.0
Dicyandiamide	2.0

Properties:

Visc. —before thickening	170 cps
—after thickening	1700 cps

Procedure:

Add ingredients in order listed. Add the dicyandiamide slowly and continue stirring for 1 h. Separation starts to take place after 3 days. Mild agitation redisperses the system.

No. 2

Dimethyl Formamide	45.0
Methyl Ethyl Ketone	45.0
Epon® 828	25.0
Carbopol® 934	3.0
Di-2(Ethylhexyl) Amine	7.0
Methanol	10.0

Properties:

Visc. —before thickening	50 cps
—after thickening	56,000 cps

Procedure:

Add ingredients in order listed. With moderate agitation, disperse the Carbopol® resin slowly to avoid lumping. Add the amine rapidly and after 10 min add the methanol. Thickening will occur with the addition of methanol. The result is a Jello-like rheology.

EMULSION ADHESIVES

Water-Based Emulsion Adhesive

Formula No. 1

(Anionic, Polybutene)

Amoco® Polybutene	100.0
Oleic Acid	3.5
Triethanolamine	1.7
Water	67.0

Procedure:

Mix polybutene with oleic acid and triethanolamine in a high-speed shear mixer such as a Cowles Dissolver. Add water slowly until inversion occurs. Add remaining water more rapidly. Preheating both the polymer mixture and the water facilitates emulsification. With the higher viscosity polymers (> 175 cs @ 99 C), temperatures of 65–90 C may be necessary. Premixing the triethanolamine with eight parts of water, adding this solution to the polymer/oleic acid mixture, and then adding remaining water may also facilitate emulsification.

No. 2

(Cationic, Polybutene)

Amoco® Polybutene	100.0
Triton® X-400	20.0
Water	67.3

Procedure:

Mix polybutene with emulsifier in a high-speed shear mixer such as a Cowles Dissolver. Add water slowly until inversion occurs. Add remaining water more rapidly. Preheating both the polymer mixture and the water facilitates emulsification. With the higher viscosity polymers (> 175 cs @ 99 C), temperatures of 65–90 C may be necessary.

	No. 3	No. 4
	(Nonionic, Polybutene)	
Amoco® Polybutene	100	100
Alkyl Phenoxy Polyethoxy Ethanol	4	—
Ethoxylated Thioether	—	5
Water	29	67

Procedure:

Mix polybutene with emulsifier in a high-speed shear mixer such as a Cowles Dissolver. Add water slowly until inversion occurs. Add remaining water more rapidly. Preheating both the polymer mixture and the water faciliates emulsification. With the higher viscosity polymers (> 175 cs @ 99 C), temperatures of 65–90 C may be necessary.

	No. 5	No. 6	No. 7	No. 8
	(Polybutene/Resin)			
Base Emulsion:				
Amoco® Resin 18-210	40	40	40	40
Amoco® Polybutene H-100	40	40	40	40
Thermoplastic Elastomer	100	100	100	100
Aromatic Solvent	100	100	100	100
Emulsifiers	8	8	8	8
Antioxidant	2	2	2	2
Water	90	90	90	90
Tackifier:				
Resin/Hexane Sol'n. (3/1)	380	380	380	380
Acrylic Latex	0	63	126	252

Procedure:

Heat **Resin 18** in solvent at 80 C until dissolved. Add preheated (80 C) polybutene and stir. Pour solution over TPE and heat 15 min in oven at 99 C. Place on pebble mill for 2–3 h. Alternately heat and roll until homogeneous. Heat to 80 C and mix in high-speed stirrer at 2000 rpm. Gradually add preheated (80 C) emulsifiers and antioxidant. Add boiling water slowly until inversion occurs. Add remaining water. Mix for 10 min. Blend in tackifier and latex.

No. 9

(Polyvinyl Acetate)

Polyvinyl Acetate Emulsion (55% solids)	100.0
Benzoflex 2-45	10–50.0
Clay Filler	0–30.0
Cooked Starch or Dextrin	0–1000.0
Sodium Benzoate (tech.)	0–2.0
Stabilizer	0–2.0
Wetting Agent	0–0.2
Water (secondary)	0–100.0
Defoamer	0–2.0
Methyl Salicylate	0–1.0

Plasticizer-Based Emulsion Adhesive

	No. 1	No. 2
	(Polyvinyl Acetate)	
	Wet	*Dry*
Polyvinyl Acetate Emulsion	100	100
Plasticizer	11	20

No. 3

(Polyvinyl Acetate)

Airflex 400	100
Plasticizer*	0–20

---

* Dibutyl phthalate, **Benzoflex 9-88** or **KP 140.**

Uses:
    Pigmented, plasticized polyvinyl chloride.

No. 4

(Polyvinyl Acetate)

Airflex 400	100
Dibutyl Phthalate	10

Uses:
Clear vinyl chloride-vinyl acetate copolymer.

No. 5

(Polyvinyl Acetate)

Airflex 400	100
Plasticizer*	10

---

* **Santicizer 8, Santicizer 141** or diisobutyl phthalate.

Uses:
Cellulose acetate and Saran-coated cellophane.

No. 6

(Polyvinyl Acetate)

Airflex 400	100
Toluene	5–10

No. 7

(Polyvinyl Acetate)

Airflex 400	75
Airflex 405	25
Dibutyl Phthalate	5

Uses:
Above two formulas good for polystyrene.

No. 8

(Polyvinyl Acetate)

Airflex 400	50
Flexbond 150	50
Dibutyl Phthalate	10–15

Uses:
Mylar.

No. 9

(Polyvinyl Acetate)

Airflex 400	75
Airflex 405	25
Dibutyl Phthalate	5–10

No. 10

(Polyvinyl Acetate)

Airflex 400	100
Dibutyl Phthalate	5–10

Uses:
Above two formulas good for annealed aluminum foil.

No. 11

(Polyvinyl Acetate)

Airflex 400	75
Airflex 410	25

Note:
Useful as a sprayable adhesive.

Resin Emulsion Adhesive

	Formula No. 1	No. 2
	(Vinyl/Resin)	
Vinsol®	100	100
Polyvinyl Alcohol	73	63
Dibutyl Phthalate	31	25
Polyvinyl Alcohol	1	1
Toluene	35	25
Water	165	140

Procedure:

Emulsify the polyvinyl acetate in an equal amount of water containing the polyvinyl alcohol as an emulsifying agent. Heat the **Vinsol®** with the dibutyl phthalate and toluene at 55–60 C to form a smooth solution (10–15 min). This **Vinsol®** solution is then slowly added in alternate increments with the remaining water to the stirred polyvinyl acetate emulsion. When using a mechanical stirrer at 3000–12,000 rpm, time required for adding **Vinsol®** solution is usually about 10–15 min. During this time interval, the stirrer speed should be increased as the emulsion thickens. Stable emulsions of 45–65% solids are readily obtained.

No. 3

(Cellulose/Resin)

Ethyl Cellulose N-50	20
Staybelite®	60
Castor Oil	20

Note:

Depending on the temperature of the hot-melt and the length of time the mix is held at this temperature, an antioxidant, such as octylphenol, and an acid acceptor, such as **Epi-Rez 510**, should be added in varying amounts between 0.25–1.0% of the ethyl cellulose content.

The above general formula may be modified in many ways to meet the needs of specialty applications. **Vinsol®** resin may be substituted for the **Staybelite®** resin at lower cost when color is not an important factor. A few of the specialty applications are:

> High-speed thermoplastic packaging
> Laminating cellophane or foils to paper
> Adhering films and artificial leather to metal
> Holding metal parts together during welding.

TEXTILE ADHESIVES

Elastomer Bonds for Textiles

Formula No. 1

(Natural Rubber)

Natural Rubber (SMR 5)	100.0
Zinc Oxide	3.5
Stearic Acid	1.5
GPF Black	35.0
Process Oil	3.0
Permanax BLW	1.5
Vulcafor CBS	1.0
Sulfur	2.5

Press cure: 20 min at 150 C.

No. 2

(SBR)

SBR 1500	100.0
Zinc Oxide	5.0
Stearic Acid	1.0
GPF Black	35.0
Process Oil	5.0
Permanax BLW	1.5
Vulcafor CBS	1.2
Sulfur	1.8

Press cure: 35 min at 150 C.

No. 3

(Butyl Rubber)

Polysar Butyl 301	100.0
Zinc Oxide	5.0
Stearic Acid	1.0

GPF Black	50.0
Process Oil	10.0
Vulcafor MBT	0.5
Vulcafor TMTD	1.0
Sulfur	2.0

Press cure: 30 min at 160 C.

No. 4

(Polychloroprene)

Polychloroprene (W type)	100.0
Light Calcined Magnesia	4.0
Zinc Oxide	5.0
GPF Black	35.0
Naphthenic Process Oil	5.0
Permanax OD	1.5
Ethylene Thiourea	0.5

Press cure: 30 min at 150 C.

Elastomer-to-Rayon Adhesive

(Water-Based, Latex)

Penacolite Resin R-2170 (65%)	26.7
Deionized Water	540.5
Sodium Hydroxide (10%)	8.0
Formalin (37%)	13.5
SBR Latex (40%)* (preferably cold rubber type)	200.0
Vinyl Pyridine Type Latex (40%)	50.0

---

* May be partially (10–20%) replaced by natural rubber latex.

Procedure:

Successively add the caustic soda and **Penacolite Resin R-2170** solutions to half of the water and then add the formaldehyde. This solution should be maintained at 70–80 F. The remaining water is added to the mixed latices. The resulting resin solution is then added to the diluted latex with agitation. The dip may be used immediately, but adhesion improves with aging. Adhesives for natural or SBR stocks are normally aged for 2–24 h.

Elastomer-to-Nylon Adhesive

(Water-Based, Latex)

Penacolite Resin R-2170 (65%)	26.7
Deionized Water	407.4
Sodium Hydroxide (10%)	8.0
Formalin (37%)	20.3
Vinyl Pyridine Type Latex (40%)	250.0

Procedure:

Successively add the caustic soda and **Penacolite Resin R-2170** solutions to half of the water and then add the formaldehyde. This solution should be maintained at 70–80 F. The remaining water is added to the mixed latices. The resulting resin solution is then added to the diluted latex with agitation. The dip may be used immediately, but adhesion improves with aging. Adhesives for natural or SBR stocks are normally aged for 2–24 h.

Flannel Foam Laminate Adhesive

(Latex)

Hycar® 2671	100
Titanium Dioxide (50% disp.)	5
Carbopol® 934	1
Ammonium Hydroxide (10% sol'n.)	to pH 8

Procedure:

Seal-coat the backside of the flannel with a continuous film of latex. Prepare a 5% stock solution of **Carbopol® 934** resin and add the latex when the solution is homogeneous. Neutralize to pH 8 with ammonium hydroxide. Stretch the flannel cloth, with backside up, over a glass plate. Apply a portion of the thickened latex over the entire area as a 15-mil coating. Cure for 10 min at 250 F. Whip the remaining thickened latex in a Hobart Mixer equipped with a wire beater. After 5–8 min of high-speed agitation, the emulsion will expand to approximately 500% of the original volume. Pour the whipped latex over the cured seal-coat of the flannel. Set the doctor blade so as to coat the fabric with a 25-mil coating. Place a polyurethane foam slab quickly over the latex and hold fast for 10 s with a glass plate. Cure the assembly 10 min at 250 F.

Polyester-to-Flannel Adhesive

(Latex)

Hycar® 2671	100
Titanium Dioxide Dispersion	5
Carbopol® 934	1
Ammonium Hydroxide	to raise pH to 8

SBR-to-Polyester Fiber Adhesive

(Styrene-Butadiene/Resin)

Ameripol 1011	80.0
Ameripol 1013	20.0
Dymerex	30.0
Pexate 549	13.0
Foral® 85	30.0
Dixie Clay	20.0
Cab-O-Sil®	1.5
Zinc Oxide	10.0
AgeRite Superlite	2.0
Hexane	460.0
Toluene	51.0

Fabric-to-SBR Foam Adhesive

(Latex)

Hycar® 2679X6	100.0
Acrysol® ASE-60	0.5
Ammonium Hydroxide	adjust to pH 6.3

Cloth-to-Polyurethane Film Adhesive

	Formula No. 1	No. 2	No. 3
	(Acrylic Latex/Resin)		
Geon® 460X1 (51% solids)	100.00	—	—
Hycar® 2679 (48.5% solids)	—	100.000	—
Hycar® 1572X45 (47.7% solids)	—	—	100.00
Amino Methyl Propanol (100% solids)	1.00	1.000	1.00
Colloid 60 (100% solids)	0.25	0.250	0.25
N,N-Dimethyl Formamide (100% solids)	2.00	2.000	2.00
Carbopol® 934 (5% solids)	1.25	0.625	1.88
Ammonium Hydroxide (28% solids)	to pH 8.6	Yes	Yes

Cure 3 min at 300 F.

Fiber-Glass Cloth Adhesive

(Acrylic)

Potassium Silicone	10
Water	5

Procedure:

Mix; to this add slowly with mixing Acrysol® 60 until you get thickening desired; then apply.

Burlap-to-Paper Adhesive

(Water-Based, Latex)

Hycar® 2570X1 (56%)	180
Tamol® 731 (25%)	8
No. 10 Whiting	150
Alcogum AN 10 (5%)	40
Water	150

Mastic Adhesive and Barrier Coating for Cotton Batting

(Latex)

Ucar® Latex 165	46.37
Tergitol® NP-44	0.83
Calgon T	1.11
Tricresyl Phosphate	5.10
Polypropylene Glycol	0.09
Atomite®	23.18
Duramite	23.18
Dowicil 5-13 Preservative	0.14

Silk Screen Adhesive

(Water-Based, Starch)

Solvitose H is dissolved in the proportion of 1:2.5, of cold water, while stirring. Stirring is continued until a homogeneous solution is obtained. Varying with the fabric, this solution may be diluted further with cold water.

Uses:

For sticking fabrics on screen printing tables or on the conveyor belts of screen printing machines.

Warp Size for Denims

(Starch)

Pearl Starch (to 106 gal finished size)	80 lb
Softener (tallow, sulfonated oil, calcium chloride)	15 lb
Kerosene	2 pt

Procedure:

Cook to 190 F and homogenize @ 2000 psi.

Warp Size for Sateen

(Starch)

Starch (to 225 gal)	160 lb
Special Softener	5 lb
Kerosene	1 pt

Procedure:
Cook to 200 F and homogenize @ 4000 psi.

Note:
The laundry industry is another substantial consumer of starch—specifically heavy-boiling starch—in which its adhesive properties combine it to the textile fibers, adding stiffening and soil protection, effectively, attractively, and inexpensively. Wheat starch has been preeminent in the laundering of men's shirts. Numerous special laundry starches are produced for this purpose, used both in liquid state and in cold-water dispersible form.

Textile-Backed Floor Covering Adhesive

(Polyvinyl Ether/Resin)

Lutonal A 50 (ca. 70% in ethanol)	6
Balsam Resin WW	25
Ethanol	10
Toluene	1
Chalk (small particle size)	58

Procedure:
First produce a homogeneous solution of **Lutonal A 50** and **Balsam Resin WW** in ethanol/toluene in a dissolver. Then add the filler portion-wise with stirring and stir until a smooth homogeneous paste is obtained.

Aluminum-to-Canvas Adhesive

	Formula No. 1	No. 2
	(Ethylene Dichloride/Resin)	
Thiokol FA	1000	1000
Benzothiazyl Disulfide	3	3
Diphenyl Guanidine	1	1
Zinc Oxide	100	100
Sulfur	15	15
Vinsol®	—	250
Ethylene Dichloride	4470	5480

Properties:

Aluminum-to-Canvas Stripping Strength*	Bond strength	
Without gasoline soaking, lb/in.	11	10
After 24-h gasoline soaking, lb/in.	10	11

---

* Bonding conditions: One adhesive coat is applied to the aluminum panel with brush and three adhesives coats to No. 10 canvas. Before they are completely dry, bond is made by rolling with a 2-lb hand roller. Bonds are dried for 3 days at 122 F before testing.

Carpet Backing Adhesive

Formula	No. 1	No. 2	No. 3	No. 4
Carboxylated SBR	100.0	100.0	100.0	100.0
Filler	400.0	400.0	400.0	400.0
Dispersant	1.0	1.0	1.0	1.0
Foaming Agent	0.3	0.3	0.3	0.3
Thickener	0.6	1.8	1.8	1.8
Indopol Polybutene H-25				
(50% solids emulsion, nonionic)	—	10	—	—
Indopol Polybutene H-100				
(50% solids emulsion, nonionic)	—	—	10	—
Indopol Polybutene H-300				
(50% solids emulsion, nonionic)	—	—	—	10

Procedure:

Slowly mix the SBR latex, dispersant, foaming agent, and a nonionic polybutene emulsion. The filler is added while mixing, and the thickener is added to bring the viscosity into the range of 16,000–20,000 cps. The mixture is frothed using high speed, approximately 5 min.

Properties:
180° Peel Adhesion*

Jute, N	25.8	47.6	48.5	47.1
Jute, lbf	5.8	10.7	10.9	10.6
Polypropylene, N	22.2	24.5	28.0	31.1
Polypropylene, lbf	5.0	5.5	6.3	7.0
Loop Lock Pull				
Jute, N	10.2	7.6	6.2	7.6
Jute, lbf	2.3	1.7	1.4	1.7
Polypropylene, N	8.9	6.7	8.5	10.7
Polypropylene, lbf	2.0	1.5	1.9	2.4

* Coating weight 32 oz/yd² (1086 g/m²), 5 cm (2 in.) wide specimens.

No. 5

(Polyvinyl Chloride)

Genflow 7007	40.0
Water	13.2
Chlorowax 500	5.0
Calcium Carbonate	40.0
Thickener	1.2
Thermoguard S	5.0

Tufted Carpet Anchor Coat

Formula	No. 1	No. 2	No. 3	No. 4	No. 5	No. 6
		(Lick-Roll Application)			(Foam Application)	
NR Latex						
(LA-SPP type, 60%)	167.0	167.0	167.0	167.0	167.0	167.0
Heveaplus MG49						
Latex (50%)	—	—	60.0	200.0	30.0	30.0
Glofoam HE (25%)	3.0	3.0	3.0	3.0	—	—

	No. 1	No. 2	No. 3	No. 4	No. 5	No. 6
				(cont'd.)		
Ammonium Oleate (20% sol'n.)	—	—	—	—	5.0	5.0
Tetrasodium Pyro- phosphate	1.0	1.0	1.0	1.0	—	1.0
Water (to 72–75% total solids content)				as necessary		
Filler	400.0	100.0	400.0	100.0	—	50.0
Thiourea (10% sol'n.)	10.0	10.0	10.0	10.0	10.0	10.0
Flexzone 6H (50%)	2.0	2.0	2.0	2.0	2.0	2.0
Antifoam Agent				as necessary		
Rohagit S-MV (10%)	4.0	4.0	4.0	4.0	4.0	4.0
				or as required		

Procedure:

The latex is mixed with the stabilizer (and, where specified, the **Heveaplus MG49** latex) and the pyrophosphate and water added. The filler is added slowly to the stabilized diluted latex to avoid localized de-hydration and consequent coagulation. The remaining ingredients are added in the order given in the formulations, the stirring continued until the mix is homogeneous. Minor adjustment of the thickener dosage may be required to obtain the desired viscosity and, in the case of lick-roll appli-cation, this should be done after the air entrained during mixing has been allowed to rise to the surface.

The mechanical working of the mix should be restricted to that neces-sary to disperse the filler, since excessive high speed stirring may result in decreased stability.

Mixes prepared from formulations 5 and 6 must be strained before use in the foaming equipment.

Polypropylene Carpet-to-Plywood Adhesive

(Latex/Resin)

Hycar® 1562X103	1250
Geon® 450X20	500
Picco® A60	750
Acrysol® GS	25
Toluene	125
Dixie Clay Dispersion	750

Laminating Cement for Foam Carpet Backing

(Styrene-Butadiene/Resin)

Solprene 300	100
Dymerex	75
Stanwhite Whiting	50
Paragon	50
Antioxidant 2246	2
Rubber Solvent Naphtha	400

Urethane Carpet Underlay

(Polyurethane)

Voranol CP-3000	100.00
Silicone L-540	12.50
Polycat 12	0.30
Stannous Octoate	0.25
Water	2.50
TDI (80/20) (index)	105.00

Properties:
Cream Time	18 s
Rise Time	150 s

Fabric Coating

(Polyvinyl Chloride)

PVC	100
Chlorowax 70	15
Thermoguard S	5
Toluene	355
Methyl Ethyl Ketone	45

Fabric Laminating Adhesive

Formula No. 1

(Acrylic Latex/Resin)

Hycar® 2600X138 (50% T.S.)	90.30
Carbopol® 934	0.24
Dioctyl Phthalate (100%)	9.00
Sodium Hydroxide (10%)	0.46

Procedure:

Wet out the **Carbopol®** in the plasticizer and add the mixture to the latex using high-speed mixing equipment. Add the caustic while agitating.

No. 2

Hycar® 2600X138 (50% T.S.)	84.1
Hycar® 2679 (48% T.S.)	8.8
Carbopol® 960 (5% sol'n.)	7.1

Procedure:

To the blend of the latexes add the **Carbopol** while agitating.

No. 3

(Acrylic Resin)

Rhoplex E-32	89.75
Defoamer	0.25
Acrysol® ASE-60	3.00
Water	3.00
DAP (35%)	4.00

No. 4

(Acrylic Resin)

Rhoplex E-358	92.75
Defoamer	0.25
Catalyst A (25%)	2.00

Acrysol® ASE-60	3.00
Acrysol® ASE-95	1.00
Ammonium Hydroxide (28%)	1.00

No. 5

(Acrylic Resin)

Rhoplex E-269	99.30
Defoamer	0.25
Ammonium Hydroxide (28%)	0.45

Fabric-to-Fabric Adhesive

Formula No. 1

(Knife-Coating Compound, Latex/Resin)

Hycar® 2671	100.0
Carbopol® 934	2.0
Ammonium Hydroxide to raise pH to 6.0–7.0	

No. 2

(Spray-Coating Compound, Latex/Resin)

Hycar® 2671	100.0
Carbopol® 934	0.1
Ammonium Hydroxide to raise pH to 6.0–7.0	

Laminates should be dried at 107–121 C (225–250 F) and cured for 3 min at 149 C (300 F).

No. 3

(Acrylic Resin)

| **Rhoplex E-358** | 95.0 |
| **Nopco® DF-160L** (diluted 1:1 with water) | 0.3 |

Ammonium Nitrate (25% aq. sol'n.)	1.9
Acrysol® ASE-60	2.8
Ammonium Hydroxide (28% NH_3)	to pH 7.9–8.3

Properties:

Visc. (Brookfield LVF #6/10 rpm)	50,000 cps

Procedure:

The catalyst is first added to the **Rhoplex** emulsion to reduce initial viscosity and thus facilitate the addition of the undiluted thickener. **Acrysol® ASE-60** is a pH-dependent thickener. When neutralized with ammonium hydroxide this acid-containing, acrylic emulsion copolymer swells greatly, clarifies and becomes highly viscous. After drying, a curing time of 1–3 min at 300 F should be sufficient. Actual time and temperature requirements will depend on the fabrics being treated and the type of curing used.

No. 4

(Dip Compounds, Resin)

Fabric dip compounds are prepared in two steps. First, prepare a 6.5% resorcinol-formaldehyde resin masterbatch with the following recipe and procedure.

Resorcinol	4.1
Formaldehyde (37%)	6.0
Sodium Hydroxide	0.1
Soft Water	89.8

pH should be 7.0

Procedure:

Dissolve sodium hydroxide in the water. Dissolve resorcinol in sodium hydroxide modified water. Add formaldehyde and mix 5 min. Let stand 2 h at 75–78 F before combining with latex.

To prepare the latex-resorcinol-formaldehyde fabric treatment compound, use the following typical recipe and procedure.

(Latex/Resin)

Hycar® 1571 (41% solids)	83.9
Resin Masterbatch (above)	91.5
Soft Water	20.6
Ammonium Hydroxide (28%)*	10.0

Procedure:

Add latex to water. Add resin masterbatch slowly to mixture with constant slow agitation for 5 min. Add ammonium hydroxide to adjust pH to 10.3. Continue agitation at least 10 min. Let compounded dip stand 16–18 h at 75–78 F before using. Maximum shelf life is 96 h at 75–78 F

---

* Variable—adjust to pH 10.3

Use:

Apply the latex-resorcinol-formaldehyde compound to a fabric by common saturating techniques. Cure from 1–5 min at 285–350 F. Moisture content of the fabric should be less than 2% after drying.

Fabric Bonding Adhesive

Formula No. 1

(Latex)

Hycar® 2679	48
Oxalic Acid	10
Methocel MC (4000 std)	4

Note:

To avoid adhesive strike-through and to maintain proper hand and adhesion, thicken the latexes and control the solids of the compound. Brookfield viscosities of 15,00–30,000 cP at 20 rpm may be obtained with either cellulosic or polyacrylic thickeners. To gain greater adhesive durability, without added catalysts, heat the dried latex adhesive for 3–5 min at 310 F for full curing, or crosslinking. Adding a catalyst will lower time–temperature cures. Oxalic acid or citric acid catalysts may be used in formulations containing no pH-dependent thickeners or thermosetting crosslinking resins.

No. 2

(Latex/Resin)

Hycar® 2671	53
Melamine Formaldehyde Resin	25
Diammonium Acid Phosphate	10
Carbopol® 934	5
Ammonium Hydroxide	28

Note:

Cures are usually obtained at 260 F. If undercured, maximum properties will develop after several days aging. Even greater durability may be achieved by combining a thermosetting resin with the reactive latex polymer. Such resins have limited stability to low pH, so higher pH and latent-acid catalysts are used for compound stability.

FLOCKING ADHESIVES

General-Purpose Flocking Adhesives

Formula No. 1

(Latex)

Hycar® 2671 (53% T.S.)	100.0
Carbopol® 940 (10% suspension in perchlorethylene)	2.0
Ammonium Hydroxide	4.0
Thermosetting Resin	2.0
Antifoam	0.2

Properties:

pH	7–8
Visc. @ 20 rpm LVT	20,000 cps

Procedure:

Agitate the latex moderately at low speed. Avoid cavitating to prevent air bubbles. Add the 10% **Carbopol® 940** solution rapidly. Agitate until homogeneous (15–20 min). Neutralize with ammonium hyroxide to desired pH. Add the thermosetting resin, then the antifoam agent. Agitate until homogeneous.

Note:

When a self-thickening latex such as **Hycar® 2671X13** is used, the level of **Carbopol®** resin needed will be lower.

	No. 2	No. 3
	(Low-Temperature Flexibility, Latex)	
Hycar® 2671	100.0	—
Hycar® 2600X92	—	100.00
Carbopol® 934	0.5	0.25
Ammonium Hydroxide	q.s.	q.s.

Application:

Knife-coat the cotton print cloth with the latex compound to provide a barrier coat before applying the adhesive coat. Apply a 10-mil wet latex adhesive coat. Vibrate the sample on a beater bar and dust with 0.30-in. rayon flock. Cure the samples for 15 min at 135 C (275 F).

No. 4

(Latex)

NR Latex (60%)	166.7
Potassium Hydroxide (10% sol'n.)	4.0
Potassium Caprylate (20% sol'n.)	2.5
Sulfur (50% dispersion)	2.0
ZDC (50% dispersion)	2.0
Zinc Oxide (50% dispersion)	2.0
Antioxidant (50% dispersion)	2.0
Thickener Sol'n.*	

---

* as required to give a viscosity of approximately 500 cP (Brookfield LVT 12 rpm).

Note:

The addition of tackifying resins to the "adhesive" compound will not improve the results if the process is carried out as specified above. If, however, the flock is applied to a gelled layer of latex, then a tackifying resin may improve the adhesion by increasing the natural stickiness of the wet gel. None the less it is unlikely that this approach would give as good adhesion as the method previously suggested.

No. 5

(Water-Based, Acrylic Latex)

Water	19.61
Ucar® Latex 878	73.36
Cellosize® QP-52,000-H } Premix	0.60
Flexol® Plasticizer TOF	1.83
Polypropylene Glycol	0.07
Oxalic Acid (10% aq. sol'n.)	2.04
Ammonium Hydroxide (28% aq. sol'n.)	1.01
Paragum 109	0.75
Cymel® 301	0.73

Procedure:
Dilute **Ucar® Latex 878** in water.   Disperse **Cellosize®** HEC in **Flexol® Plasticizer TOF** and add to **Ucar® Latex 878.**   Add polypropylene glycol to control foaming.   Add the oxalic acid solution and allow the mixture to blend until smooth (about 20–30 min).   Then add the ammonium hydroxide, **Paragum 109,** and **Cymel® 301** in the order listed.

No. 6

(Water-Based, Acrylic Latex)

Rhoplex® E-358	1000.0
Nopco® DF-160L (diluted 1/1 with water)	3.0
Water ⎤ Slurry	37.8
Methocel® J-12-HS ⎦	5.4
Ammonium Hydroxide (28–30% NH_3)*	0.3
Oxalic Acid (10% aq. sol'n.)**	25.0

*	pH	8.3
	Visc. (Model LVF Brookfield #4 spindle, 6 rpm)	62,000 cps
**	pH	2.3
	Visc. (Model LVF Brookfield #4 spindle, 6 rpm)	50,000 cps

No. 7

(Acrylic)

Rhoplex® E-358	1000.0
Nopco® DF-160L (diluted 1/1 with water)	3.0
Catalyst A (25% aq. sol'n.)	20.0
Acrysol® ASE-60	29.0
Ammonium Hydroxide (28–30% NH_3)	to pH 8–8.5

Properties:
Visc. (Model RVF Brookfield #6 spindle, 10 rpm)	50,000 cps

No. 8

(Acrylic Resin)

Rhoplex® E-32	81.2
Nopco® DF-160L (diluted 1/1 with water)	0.3
Acrysol® ASE-60 ⎫	3.3
Acrysol® ASE-95 ⎬ Premix	1.6
Water ⎭	4.9
Curing Agent RK-8	8.1
Ammonium Hydroxide (28–30% NH_3)	0.6
	(to pH 8.0)

Properties:
Visc. (#4 spindle, 6 rpm)

as formulated	60,000 cps
after 5 days	58,000 cps

No. 9

(Acrylic Resin)

Rhoplex® TR-520	85.90
Flexol® TOF	3.40
Nopco® DF-160L (diluted 1:1 with water)	0.15
Cellosize® QP-300 ⎫ Premix	0.90
Water (cold) ⎭	8.60
Ammonium Hydroxide (28% NH_3)	1.10

Properties:

pH	7.7
Visc. (#4 spindle, 6 rpm)	70,000 cps

Application:
This composition is applied by knife/roll coater to a cotton fabric and rayon flock is applied. The fabric is cured at 300 F for about 4 min.

Nylon Flock-to-Steel Adhesives

Formula No. 1

(Acrylic/Resin)

Hycar® 2600X172	48.2
CMC 12H	5.0
Acrysol® ASE-75	40.0
Phosphoric Acid	10.0
Ammonium Hydroxide	28.0

Properties:

Total compounded solids	44.3%
Visc. (Brookfield RVF @ 20 rpm)	8000 cP

Application:

0.3 mm (12 mil) (wet) adhesive coat, dry 10 min @ 100 C (212 F) and cure 15 min @ 135 C (275 F).

No. 2

(Acrylic/Resin)

Hycar® 2600X172	100.00
CMC 12H	0.81
Acrysol® ASE-75	1.00
Phosphoric Acid	0.35
Ammonium Hydroxide	to pH 6.4

Flock-to-Polyurethane Foam Adhesive

	Formula No. 1	No. 2
	(Acrylic/Resin)	
Hycar® 2671	100.00	100.00
Cyrez 933	—	3.00
Carbopol® 934	0.75	0.75
Ammonium Hydroxide	to adjust pH to 9.0	

Dry at 100 C (212 F) for 15 min and cure at 135–149 C (275–300 F) for 15 min. Formulations are thickened at a Brookfield visc. RVF (20 rpm) of 10,000 mPa·s (cP).

No. 3

(Acrylic/Resin)

Hycar® 1572X45	100.00
Cyrez 933	3.00
Carbopol® 934	0.75
Ammonium Hydroxide	to pH 9.0

No. 4

(Acrylic/Resin)

Acronal® 330 D	100.0
Nopco® NXZ	0.1
Collacral® VL	2.0

Note:

After drying, the adhesive film can be cross-linked by heating for 5 min @ 140 C. Adhesives made up to this formulation will have a suitable consistency for blade or roller application. Polyurethane foam packaging inserts are sprayed with adhesive. **Propiofan® 5 D** can also be used for this purpose. The viscosity for spraying is best adjusted by incorporating **Propiofan® 6 D.**

Foam Flock Adhesive

(Acrylic)

Rhoplex® K-87	81.9
Melamine Resin	1.6
Water	3.8
Acrysol® ASE-60 } Premix	4.5
Acrysol® ASE-95	3.1
Ammonium Nitrate (25%)	1.6
Ammonium Hydroxide (28%)	0.6
Ammonium Stearate (33%)	2.9

Properties:

pH (adjusted to)	8.0
Visc. (#4 spindle, 6 rpm)	50,000 cps

Note:

This formulation should be foamed to a density of 0.5–0.6 g/cm^3 before applying to the fabric substrate. After flocking, the foamed adhesive is dried at a temperature of 240 F (116 C) and then cured at 300 F (149 C).

Other additives such as opacifiers and optical brighteners may be added to the formulation if desirable. In addition, the use of a blocked melamine extends the pot life of the formulation.

Flock-to-Vinyl Adhesive

	Formula No. 1	No. 2	No. 3
		(Latex)	
Geon® 460X1	100.00	100.0	—
Geon® 576	—	—	60.00
Geon® 351	—	—	40.00
Tetrasodium Pyrophosphate	0.10	—	—
Isophorone	—	2.0	—
Cymel® 300	—	—	5.00
Carbopol® 934	1.26	0.5	—
Trichloroethylene	5.00	—	—
Methocel® HG-65	—	—	0.75
Properties:			
pH	7.2*	5.5	7.1
Visc. @ 20 rpm, cP	28,000	2900	1560

* pH adjusted with 28% ammonium hydroxide.

Flock-to-Fabric Adhesive

(Acrylic Resin)

A*	Acronal 35 D	100.0
	Nopco® NXZ	0.1
	Latekoll® D	4.0
B	Kaurit® M 70	

* adjusted to pH 6 with concentrated ammonia

Procedure:
Mix 100 parts of Part A with 2–6 parts of Part B prior to use.

Note:
The made-up adhesive has a pot life of approx. 8 h. The **Kaurit® M
70** is added in order to render the flocked surface very resistant to washing
and dry cleaning, but any increase in the amount added will impair the
handle of the flocked material.
Fabrics can be flocked over the entire surface or in designs.

Flock-to-Plasticized PVC Film Adhesive

(Acrylic Resin)

Acronal 500 D	100
Collacral® VL	2

Note:
The adhesive can be applied by blade or roller. For spray application,
the proportion of **Collacral® VL** must be reduced.   The adhesive film
has very good resistance to plasticizers.
Flocked plasticized PVC films are also used for decorative purposes.

Flock-to-Rigid PVC and Polystyrene Adhesive

(Acrylic Resin)

Acronal 290 D	600.0
Acronal 500 D	255.0
Palatinol® C	18.0
Nopco® NXZ	1.7
Resin Sol'n.*	85.3
Water	40.0

---

* The resin sol'n. consists of:

Soft Resin KTN	70
Toluene	15
Butyl Acetate	15

Note:
    An adhesive made up to this suggested formulation yields flocked sheet
that can be thermoformed to produce such articles as plastics packaging

inserts. This adhesive can be applied by blade. For spraying, it must first be diluted with water. When dried, the adhesive film has adequate abrasion resistance.

Flock-to-Pretreated Polyethylene Adhesive

(Acrylic Resin)

Acronal 500 D	60
Epotal® 181 D	40
Collacral® VL	2

Note:
Polyethylene film can be blade-coated with this adhesive. For flocking polyethylene flowers, the adhesive is applied by dipping or spraying; it must first be diluted with water.

Flock-to-Paper, Board, and Wall Coverings Adhesive

(Polyvinyl Propionate)

Propiofan® 5 D	100
Methyl Cellulose (4% sol'n.)	20

Note:
A 2% aqueous solution of the methyl cellulose used had a viscosity of around 600 mPa·s (Höppler, 20 C). Methyl cellulose yields the long open-assembly time required for flocking. The adhesive can be applied by blade or roller.

Flock-to-Fiberboard, Hardboard, and Wood Adhesive

(Urea Resin)

Urecoll® 118	100
Ammonium Chloride (15% sol'n.)	10

Note:

This adhesive has a pot life of approximately 10.5 h @ 20 C and about 3.5 h @ 30 C. After curing, the flocked coating can be cleaned by wiping with a damp cloth. **Urecoll® 118** is a urea resin solution that yields a hard film.

The viscosity allows for roller application.

AUTOMOTIVE ADHESIVES AND SEALANTS

General-Purpose Automotive Adhesive

(Polychloroprene, Solvent-Based)

Rubber Mill Mix:

Polychloroprene	100
Antioxidant	2
Magnesium Oxide	4
Zinc Oxide	5

Prereacted Resin Mix:

Heat-Reactive Phenolic Resin	45
Solvent: Hexane/Cyclohexane (50/50 by vol.)	408.3 1
Magnesium Oxide	4
Water	1

Final Mix:

Prereacted Resin Mix	total mix
Rubber Mill Mix	total mix
Solvent: MEK/Acetone/Ethyl Acetate (50/12.5/37.5 by vol.)	269.2 1
Amoco® Resin 18-290	30

Procedure:

Using a two-roll tile mill, the rubber mill mix is prepared, banded, and chopped into small pieces. The heat-reactive phenolic resin was prereacted in the hexane, cyclohexane, magnesium oxide, and water for 24 h under agitation. The rubber mill mix and the prereacted resin mix are combined with the remaining ingredients, and the batch is agitated until completely dissolved.

Properties:

180° Peel Adhesion	kN/m	lb/in.
Aging conditions:		
24 h @ 23 C (73 F)	11.8	67.4
3 days @ 50 C (122 F)	10.5	60.0
5 days @ 50 C (122 F)	11.0	62.8
7 days @ 50 C (122 F)	11.4	65.3

Uses:

Cabinets, countertops, sandwich panels, and various automotive applications.

Windshield Adhesive

(Polybutene/Resin)

Ethylene Propylene Terpolymer	40.40
Amoco® Polybutene H-300	40.40
Amoco® Resin 18-210	8.08
Zinc Oxide	2.02
Stearic Acid	0.41
Tetraethyl Thiuram Disulfide	0.27
2-Mercapto Benzothiazole	0.13
Sulfur	0.21
Carbon Black (FEF)	8.08

Procedure:

The EPDM is banded on a tight mill at 82–93 C (180–200 F) for 10 min. H-300 polybutene/carbon black slurry is prepared and added in increments and milled 30 min. Zinc oxide and stearic acid are added and dispersed (5–10 min). Accelerators are added and dispersed (5 min). Sulfur is added and milled for 5 min. Stock is then extruded into $1/2 \times 1/2$ in. tape. Stock is cured in a forced-draft oven for 24 h before testing.

Automotive Windshield and Backlights Sealant

	Formula No. 1	No. 2	No. 3
	(Polybutene/Butyl Polymer)		
EX-214	100	100	100
Statex R-H	90	90	90
Indopol® H-100	55	35	35
Sunpar 2100	30	30	15
ST-5115	—	—	40

Windshield Sealing Tape

Formula No. 1

(Polybutene/Butyl Rubber)

Chlorobutyl 1066	40.0
Exxon Butyl 065	60.0
Purecal U	50.0
Stearic Acid	0.4
Maglite K	0.4
Zinc Oxide	2.0
HAF-LS	100.0
Indopol® H-100	115.0

Procedure:

Heat-treat first 6 ingredients for 5 min @ 320 F in Banbury to selectively cure the **Chlorobutyl** fraction. Gradually blend the carbon black and polybutene into the masterbatch in a kneader mixer, avoiding simultaneous additions of black and polybutene.

No. 2

(Polybutene-Elastomer)

Kalar 5214	100
Statex R-H	90
Indopol® H-100	70
Sunpar 2100	30

Automotive Sealing Tape

	Formula No. 1	No. 2	No. 3	No. 4
	(Polybutene/Butyl Rubber)			
Bucar 5214	100	100	—	—
Exxon Aid-10	—	—	100	100
Amoco® Polybutene H-100	100	—	100	—
Amoco® Polybutene H-300	—	100	—	100
Stearic Acid	2	2	—	—
Super Beckacite®	—	—	20	20
Carbon Black Beads	90	90	140	140

	No. 5	No. 6	No. 7	No. 8
	(Polybutene/Butyl Rubber)			
Polysar XL-20	100	100	—	—
Polysar XL-50	—	—	100	100
Amoco® Polybutene H-100	160	—	100	—
Amoco® Polybutene H-300	—	160	—	100
Super Beckacite®	20	20	20	20
Carbon Black Beads	140	140	140	140

	No. 9	No. 10	No. 11
	(Polybutene/Butyl Rubber)		
Bucar 5214	100	100	100
Amoco® Polybutene	100	100	100
Stearic Acid	2	2	—
Super Beckacite® 2000	—	—	20
Carbon Black Beads	90	90	140

Procedure:

Time

0 min	With cold water on, charge rubber, stearic acid, $1/2$ carbon black, and $1/2$ polybutene to mixer; position ram and mix.
5 min	Add $1/4$ carbon black and $1/4$ polybutene.
10 min	Add $1/4$ carbon black and $1/8$ polybutene.
20 min	Turn off cold water and steam heat to 250 F.
30 min	Steam off, cold water on.
35 min	Add remaining polybutene.
40 min	Dump

	No. 12	No. 13
	(Polybutene/Butyl Rubber)	
Bucar 5214	100	100
Amoco® Polybutene	100	160
Super Beckacite® 2000	20	20
Carbon Black Beads	140	140

Procedure:

Time

0 min	With cold water on, charge rubber, stearic acid, $1/2$ carbon black, and $1/2$ polybutene to mixer; position ram and mix.
5 min	Add $1/4$ carbon black and $1/4$ polybutene.
10 min	Add $1/4$ carbon black and $1/8$ polybutene.
20 min	Turn off cold water and steam heat to 250 F.
30 min	Steam off, cold water on.
35 min	Add remaining polybutene.
40 min	Dump

No. 14

(Nondrying—Polybutene)

Amoco® Polybutene H-300	24.01
Amorphous Polypropylene Homopolymer	5.05
Butyl Rubber	1.72
Clay	16.37
Calcium Carbonate	44.05
Diatomaceous Silica	4.00
Cotton Fiber	4.80

Procedure:

In the laboratory the sealant is compounded in a sigma-blade mixer. Amorphous polypropylene is premixed with twice its weight of polybutene to facilitate complete dispersion. (This premixing may not be required in commercial-scale equipment.) Additions to the mixer are made in the order listed. The entire mass is then mixed for 1 h after the last addition.

No. 15

(Polyisobutylene/Polybutene)

Polyisobutylene	32.52
Oleic Acid	0.97
Amoco® Polybutene H-300	23.43
Asbestos	24.39
Asbestos	16.26
Titanium Dioxide	2.43

Procedure:

A double-arm dispersion-blade Baker-Perkins mixer is used. The polyisobutylene, oleic acid, and **H-300** polybutene are added to the mixer and allowed to mix for 15–20 min. The asbestos and titanium dioxide are then added incrementally in the order shown. A mix time of 10 min is allowed between each addition.

After adding the titanium dioxide the whole mass is mixed for 10 min. Total mix time is approximately 1 h and 15 min.

No. 16

(Polybutene/Resin)

EPDM	40.40
Amoco® Polybutene H-300	40.40
Amoco® Resin 18-210	8.08
Zinc Oxide	2.02
Stearic Acid	0.41
Tetraethyl Thiuram Disulfide	0.27
2-Mercapto Benzothiazole	0.13
Sulfur	0.21
Carbon Black (FEF)	8.08

Procedure:

The EPDM is banded on a tight mill at 82–93 C (180–200 F) for 10 min. **H-300** polybutene/carbon black slurry is prepared and added in increments and milled 30 min. Zinc oxide and stearic acid are added and dispersed (5–10 min). Accelerators are added and dispersed (5 min). Sulfur is added and milled for 5 min. Stock is then extruded into $1/2 \times 1/2$ in. tape. Stock is cured in a forced-draft oven for 24 h before testing.

No. 17

(Polybutene/Resin)

Thermoplastic Elastomer	3.84
Tackifying Resin	7.69
Amoco® Resin 18-290	9.61
Amoco® Polybutene H-1500	24.98
Calcium Carbonate	42.27

Silica Extender	9.61
Titanium Dioxide	1.92
Antioxidant	0.04
Stabilizer	0.04

	No. 18	No. 19	No. 20	No. 21	No. 22
			(Butyl Rubber/Resin)		
Polysar XL-20	100	100	80	87	58
Polysar 301	—	—	—	13	42
Royalene 400	—	—	40	—	—
N-770 (SRF)	120	70	135	100	100
N-550 (FEF)	—	50	—	—	—
Mistron Vapor®	40	—	50	40	40
Hard Clay	—	50	—	—	—
ST-5115	15	15	15	15	15
Amberol ST-149	20	20	20	20	20
Indopol® H-100	120	120	120	100	100

No. 23

(Latex)

	Dry	Wet
Dow Latex XD-8986.01 (53%)	75.0	141.5
Dow Latex 283 (45%)	25.0	56.6
Antifoamer	1.0	1.0
Calcium Carbonate	195.0	195.0
Titanium Dioxide	5.0	5.0
Ethylene Glycol	3.0	3.0
Butyl Benzyl Phthalate	10.0	10.0
Dalpad A	10.0	10.0

No. 24

(Acrylic)

Charge to a high-shear low-speed mixer (double-blade sigma or planetary type) and mix for several minutes:

Duramite	543.14
Thixatrol ST	47.73
Ti-Pure® R-901	23.59

Charge, then mix for 45 min:

Acryloid RAS-75 (@ 83% solids)	397.65
Acryloid CS-1 (@ 83% solids)	170.86

Charge, then mix for 15 min the premix:

Cobalt Naphthenate (6%)	0.60
Zinc Naphthenate (8%)	2.98
Silane A-174	1.30
Xylene	20.52
Exkin No. 2	0.50

No. 25

(Acrylic)

Charge to a high-shear, low-speed mixer (double-blade sigma or planetary type) and mix for several minutes:

Camel-Kote	543.14
Thixatrol ST	47.73
Ti-Pure® R-901	23.59

Charge, then mix for 45 min:

Acryloid RAS-75 (@ 83% solids)	397.65
Acryloid CS-1 (@ 83% solids)	170.86

Charge, then mix for 15 min the premix:

Cobalt Naphthenate (6%)	0.60
Zinc Naphthenate (8%)	2.98
Silane A-174	1.30
Xylene	20.52
Exkin No. 2	0.50

Automobile Weld-Through Seam Sealer

(Oil-Based, Styrene-Butadiene)

Solprene 1205	100
Circosol 380	250
Paragon Clay	200
HiSil 233	25
Zinc Oxide	5
Stearic Acid	3
Schenectady SP-567	20
Wingstay S	2
Bismate	2
MBTS	2
Sulfur	5
Unicel S	4
Antimony Trioxide	10
Chlorowax 50	30
Triethanolamine	5

Procedure:

Mix cycle in Baker-Perkins mixer: Soften **Solprene 1205** in hot mixer, then add clay and oil alternately to maintain good mixing. Add **HiSil**, zinc oxide, stearic acid, resin, and stabilizer. Cool batch to 150 F or below and add balance of materials.

Automotive Seam Sealer

(Oil-Based, Mineral Rubber)

A Base Grease	
Amine O	2–3
Attagel 50	10–15
Mineral Oil	82–88
B Base Grease A	60–70
Gilsonite	30–40

Procedure:

Base Grease is gelled with heat in Cowles dissolver. Cool to room temperature and place in Z-bar blender and add **Gilsonite** in proportion given. Fillers such as asbestos and ground limestone can be added in Step B to modify bead characteristics.

Automotive Hot-Melt Sealer

(Ethylene Vinyl Acetate/Butyl Rubber)

EVA-501	40
Bucar 5214	30
CK-1834	20
Santicizer 160	5
Calcium Carbonate	5

General Automotive Sealant

(Polymercaptan)

Polymer Package	
Dion 1002	100.0
Carbon Black (low oil absorption)	150.0
Ircogel 900	5.0
Catalyst	
Lead Dioxide	7.5
DOP	5.5

Radiator Sealer

(Water-Based, Dextrin)

A Water	150 gal
Starch	150 lb

Stir and heat until clear. Add to Part B.

B Water	450 gal
CMC 7HP	20 lb

Stir and mix until dissolved. Then add:

C Starch	250 lb
Dextrin	400 lb
Borax	40 lb
Asbestos	300 lb
Santobrite	18 lb
Water	206 gal

CAUTION:
Do **not** package in bimetallic containers!

Brake-Liner Cement

(Rubber/Resin)

Adhesive:

Hycar 1001	100.00
Channel Black (easy processing)	50.00
Zinc Oxide	5.00
Sulfur	3.00
Benzothiazyl Disulfide	1.25
SP-8010	90.00
Methyl Ethyl Ketone	748.50

Flexible Coating:

SP-8010	100
Polyvinyl Butyral	45
Methyl Isobutyl Ketone	400
Amyl Alcohol	400

Pressure-Sensitive Steam-Hose Adhesive

	Formula No. 1	No. 2
	(Tube)	*(Cover)*
Chlorobutyl 1066	100.00	100
HiSil 233	60.00	—
FEF	—	30
HAF	—	30
Flexon 875	—	5
Flexon 765	10.00	—

Stearic Acid	1.00	1
Amberol ST-149	3.00	3
Antioxidant 2246	1.00	1
Diethylene Glycol	2.00	—
Maglite D	0.25	2
Zinc Oxide	—	3
Litharge	10.00	—
TMTDS	—	1
MBTS	—	2
NA-22	0.75	—
Press cured @ 307 F, min	60	40

Procedure:

Time

0 min   Charge **Chlorobutyl, Maglite D, Antioxidant 2246.**

1 min   Add $^1/_2$ **HiSil**, diethylene glycol.

2 min   Add **Amberol ST-149**, stearic acid, $^2/_5$ **HiSil**.

4 min   Attain 350 F, heat-treat 6 min.

10 min   Initiate cooling water, add oil, balance of **HiSil**.

12 min   Dump at 300 F.

—   Add litharge and **NA-22** to cool masterbatch in a second Banbury pass or on cool mill.

Undertread Cement

(Rubber/Resin)

Natural Rubber	50.0
SBR 1502	50.0
FEF Black	60.0
Zinc Oxide	5.0
Stearic Acid	1.0
Antioxidant	1.5
Diphenyl Guanidine	0.3
Santocure	1.2
Sulfur	2.0
Circo Lt. Oil	10.0
CRT-336	50.0
Solvent	to 20% solids

Asphalt Cutback Underbody Coatings*

	Formula No. 1	No. 2
Asphalt Cutback	60.0	60.0
Arquad® 2C-75	1.4	1.4
Mix 2 min @ slow speed.**		
Attagel® 36	12.0	12.0
Mix 10 min @ medium speed.**		
Asphalt Cutback	13.3	13.3
Emtal® 41 or Emtal® 42 Talc	13.3	—
Wollastonite	—	13.3
7T5 Asbestos	00	0
Mix 15 min @ medium speed.**		
Clay/Surfactant Ratio	8.5/1	8.5/1
Typical Asphalt Properties:		
Softening Pt. (Ring & Ball)		175
Needle Penetration		25–26
Cutback Solids		50%/wt.
Saybolt Visc.		100 s @ 77 F

* Formulated to meet Federal Specification TT-C-520B.
** Mixer: Electrically driven laboratory paddle mixer.

Tire-Tread Splice Cement

(Polyisoprene/Resin)

Ameripol SN 600	100.00
Zinc Oxide	5.00
Stearic Acid	1.00
N-660 GPF Black	45.00
Varcum® 875	25.00
Pine Tar	5.00
AgeRite D	1.00
CBTS	1.00
TMTD	0.10
Sulfur	2.25

Procedure:
Add solvent to approximately 6% solids using 85/15 hexane/toluene blend.

Tire-Tread Cement

	Formula No. 1	No. 2
	(Styrene-Butadiene)	
Solprene 301	100.00	—
Solprene 377	—	68.80
Solprene 3001	—	50.00
Philblack N330	40.00	40.00
Zinc Oxide	4.00	4.00
Stearic Acid	1.50	1.50
Wingstay 100	1.00	1.00
Santocure	0.75	0.75
Thionex	0.20	0.20
Sulfur	2.00	2.00
Schenectady SP-1068	15.00	15.00
Rubber Solvent	to make approx. 10% solids	

Cure 24 min @ 307 F.

Automobile Headliner Cements

	Formula No. 1	No. 2
	(Styrene-Butadiene/Resin)	
Solprene 301	100	100
Dymerex	75	—
Polypale	—	50
Wood Rosin	—	70
Schenectady SP-154	30	—
Schenectady SP-1045	—	30
Hercolyn® D	5	—
Zinc Oxide	—	7
Ethyl Alcohol	25	25
Textile Spirits	567	540

Chapter VI

MISCELLANEOUS

Shoe Sole Cements

	Formula No. 1	No. 2
	(SBR/Resin)	
SBR-1500	100.0	100.00
SP-6601	15.0	15.00
HiSil 233	60.0	—
Silene EF	—	55.00
Suprex Clay	15.0	75.00
Zinc oxide	5.0	5.00
Solka Floc BW-200	5.0	5.00
Cumar® MH	8.0	8.00
Ionol	1.0	1.00
Stearic Acid	1.0	1.00
Paraffin Wax	1.0	1.00
Diethylene Glycol	3.0	—
Altax	1.5	1.50
Monex	0.3	0.25
Sulfur	2.5	2.50

No. 3

(Solvent-Based)

PA-05 or PA-06	100
Parlon P	15

| Arofene 965 | 5 |
| Solvent Blend | 480 |

No. 4

(Solvent-Based)

PA-05 or PA-06	100
Beckacite 1410	20
Solvent Blend	480

Note:

Heat resistance is not as good as Formula No. 3, but this provides easier activation.

No. 5

(Solvent-Based)

PA-20	100
Q-517	20
Solvent Blend	400

Note:

This is a good general-purpose formula that has good adhesion to many substrates; and since **Q-517** is a blocked isocyanate, it can be cured in 3 min @ 300 F.

Hot-Melt Adhesive for Shoe Counters

(Ethylene Vinyl Acetate/Resin)

Elvax 4310	28.5
Elvax 4260	19.0
Pentacite® 1031	19.0
Pentacite® 435	19.0
Microcrystalline Wax (m.p. 180 F)	14.2
Antioxidant	0.3

Properties:

Visc. @ 325 F	24,500 cps
Visc. @ 350 F	13,875 cps
Elongation ($1/_2$ in. × 20 mil film, pulled at 2 in/min)	250–300%

Durable Gasket Cement

(Solvent-Based, Rubber)

FA Polysulfide Rubber	100.0
Zinc Oxide	10.0
SRF Black	60.0
MBTS	0.4
DPG	0.1
Sulfur	0.5
Ethylene Dichloride	490.0

Note:

To make 1 gal of this cement, 2.3 lb of solids are incorporated in 6.6 lb of ethylene dichloride.

Palletizing Adhesives

Formula No. 1

(Water-Based, Starch)

Mix @ room temp.:

Water	38.8
Defoamer	0.1
Preservative	0.1
Staclipse J-UB	27.8
Star-Dri 42R	5.4

Cook 20 min @ 190–195 F. Cool below 140 F. Add:

Vinsol® Emulsion	27.8

Mix well and dilute, if required.

Properties:

Refractometer Solids	40%
pH	≈ 8
Brookfield Visc. @ 80 F	1000 cps

No. 2

(Water-Based)

Water	78.9
Preservative	0.1
Volclay 200	12.0
Koldex 30	9.0

Procedure:

Add ingredients carefully to well-agitated water at room temperature. Mix for 1 h at room temperature. Dilute, if required, to obtain desired viscosity.

Properties:

Refractometer Solids (blurred)	15–20%
pH	≈ 9
Brookfield Visc. @ 80 F	
Fresh	≈ 2500 cps
Aged	Soft thixotropic gel

Marine-Seam Compound

(Oil-Based)

Blown Marine Oil (Z3 visc.)	37.0
Raw Linseed Oil	150.0
Atomite®	385.0
Putty Filler	370.0
Mistron Vapor®	55.0
Cobalt Drier (6%)	0.5
Lead Drier (24%)	1.0
Antiskin	1.5

Uses:

A marine-seam compound is used to fill hull seams, for various bedding applications, and to seal scarf and other joints. Thus, such compounds must be water resistant, have good adhesion, set rapidly to form a tough surface yet remain sufficiently flexible to maintain a seal when dimensional changes at the seams occur.

Polyurethane Adhesive

Formula No. 1

Estane 5713	30
THF	50
Methyl Cellosolve	50
Carbopol® 941	2

Procedure:

Dissolve the **Estane 5713** in the solvent blend.   Add the **Carbopol®** resin slowly to avoid lumping.   Stir for approximately 3 h. Final Brookfield visc. is 98,000 cps @ 20 rpm.

No. 2

Estane 5703	13.0
MEK	70.0
DMF	10.0
Carbopol® 934	2.0
Di-2 (Ethylhexyl) Amine	5.0

No. 3

Estane 5713	30
THF	50
Methyl Cellosolve	50
Carbopol® 941	2

Procedure:

Dissolve **Estane 5713** in the solvent blend.   Add the **Carbopol®** resin slowly to avoid lumping.   Stir for approximately 3 h.   Final Brookfield visc. is 98,000 cps @ 20 rpm.

No. 4

Unithane 200 MC (50% solids)	50
MEK	10
Methyl Cellosolve	15
Carbopol® 941	1

Procedure:

Dissolve the **Unithane** in the solvent blend. Add the **Carbopol®** resin slowly to avoid lumping. Stir for approximately 3 h.

Note:

Visc. before addition of **Carbopol® 942** was 825 cps; after was > 1,000,000 cps.

No. 5

Estane 5713 F-1	23
DMF	50
Cellosolve	50
Carbopol® 941	2

Procedure:

Dissolve the **Estane 5713 F-1** in the solvent blend. Add the **Carbopol®** resin slowly to avoid lumping. Stir for approximately 3 h.

One-Package Vinyl/Urethane Adhesive

(Solvent-Based)

Heliol 115-3	100.0
Cymel® 300	12.5
Synasol	35.0
Toluol	25.0
VAGH	17.5
Ethyl Acetate	70.0
Mobil PA-75	4.0

Note:

Cure 1 week at room temp.

Dipping Cement

Formula No. 1

(Polyurethane/Resin)

Estane 5701	13.9
Tetrahydrofuran	37.1
Dimethylformamide	37.1
Carbopol® 934	1.6
Di-2 Ethylhexyl Amine	1.6
Dimethylsulfoxide	8.7

Procedure:

Dissolve the polyurethane in the THF-DMF blend. While agitating disperse the **Carbopol®** resin and then add the amine. The addition of DMSO will result in a Brookfield visc. (20 rpm) of 36,000 cps.

No. 2

(Styrene Foam, Latex)

Add, under agitation, one part of **Hycar® 1562X103** to 6 or more parts of **Hycar® 1312**. Because this mixture provides excellent "wet grab," it need not be thoroughly dried to realize its optimum adhesive properties.

No. 3

(Latex)

Water	—
Carbopol® 934	3.30
Sodium Hydroxide (10% sol'n.)	1.32
Ethomeen C-25	1.50
Plasticizer-Latex:	
Santicizer 160	20.00
Geon® 352 (57.3% T.S.)	100.00

Pigment Mix:

Water	—
Titanium Dioxide	15
Whiting	85

Properties:

Visc. (Brookfield, 20 rpm)	22,250 cps
pH	7.4
Total Solids (%)	40
Carbopol® 934 Concentration	0.97% on water
	0.58% on total weight

Procedure:

Carefully disperse the **Carbopol® 934** in the water and neutralize it, following the proper mixing procedure. Add the sodium hydroxide first and then the **Ethomeen C-25**. Slowly add the oils with vigorous agitation to form an oil-in-water emulsion. Slowly add the latex with gentle agitation. Add the pigment mix and stir to uniformity.

No. 4

(Solvent-Based, Vinyl)

Geon® 427	8.4
Geon® 443	8.4
Methyl Ethyl Ketone	50.3
Titanium Dioxide	22.3
Carbopol® 934	1.5
Di-2 Ethylhexyl Amine	2.6
Water	2.5
Methanol	4.0

Procedure:

Dissolve the resins in the MEK. Add the titanium dioxide and amine and ball mill to the desired pigment grind. Disperse the **Carbopol®** resin in the above blend while agitating. Blend the water and the methanol and then add the mixture to the system at which time the cement will thicken.

Nylon, Polyester-to-Rubber Adhesive

Formula No. 1

(Rubber)

Natural Rubber	80.00
Polybutadiene	20.00
Zinc Oxide	5.00
Stearic Acid	2.00
Antidegradant	3.00
Sunproofing Wax	1.00
HAF Black	25.00
Aromatic Oil	4.00
Santocure NS	1.20
Accelerator	0.10
Sulfur Bonding System	2.25
Silica	15.00
Cohedur RL	5.00

Cure 25 min (140 C).

No. 2

(Water-Based, Latex)

Penacolite R-2200 (70%)	28.6
Water (preferably deionized)	405.5
Sodium Hydroxide (10%)	8.0
Formalin (37%)	20.3
Vinyl Pyridine-Type Latex (40%)	250.0

No. 3

(Water-Based, Resin)

Water	66.2
Nopco® DF-160L (diluted 1:1 with water)	0.2
Rhoplex® TR-520	22.8
Ammonium Nitrate (25% sol'n.)	2.0
Triton® GR-5	0.5
Ammonium Hydroxide (28% NH₃)	0.9
Acrysol® ASE-60 ⎫ Premix	3.7
Water ⎭	3.7

Properties:
 Brookfield LVF Visc. (#4 spindle, 6 rpm)   ≈ 2000 cps

Rayon-to-Rubber Adhesive

(Water-Based, Latex)

Penacolite R-2200 (70%)	28.6
Water (preferably deionized)	538.6
Sodium Hydroxide (10%)	8.0
Formalin (37%)	13.5
SBR Latex (40%) (preferably cold rubber type)	200.0
Vinyl Pyridine-Type Latex (40%)	50.0

Nylon and Brass Adhesive

(Rubber)

EPDM 1440/1470	100
Silica 150	55
Naphthenic Oil	40
Zinc Oxide	15
Stearic Acid	4
Sulfur	3

Neoprene Adhesive

(Rubber/Resin)

Neoprene AH	100.0
Heptane	variable
Thiuram E	0.5
Accelerator 552	0.5
Phenolic Resin	40.0
Magnesium Oxide	2.0
Zinc Oxide	4.0
Antioxidant	2.0
Water	1.0
Tackifier	0–20.0
Thickening Agent	variable

"Duco"-Type Nitrocellulose Cement

(Solvent-Based)

Nitrocellulose (30–40 s R.S.)	10
Benzoflex 2-45	5
Ethyl Acetate	34
Acetone	34
Dicyclohexyl Phthalate	12

Noncuring Splicing and Substrate Adhesive

(Solvent-Based, Butyl/Resin)

Enjay Butyl 218	100
Piccolyte® S-125	100
Polybutene H-1900	20
Hexane	205
Toluene	205

Splice Cement

(Butyl/Resin)

Butyl 218	100.0
SP-1056	15.0
Clay	100.0
Micro-Cel	50.0
Pigment	5.0
Zinc Oxide	5.0
Polyac	0.6
Tetramethyl Thiuram Disulfide	0.1
Selenium Diethyl Dithiocarbamate	0.1

Rubber Cement

Formula No. 1

(Solvent-Based, Rubber)

Hycar® 1001X225	15
Methyl Ethyl Ketone	250

Carbopol® 934	6
Di (2-Ethylhexyl) Amine	6
Water	35

Procedure:

Dissolve the rubber in the MEK by standard procedures. Add the Carbopol® 934 with good stirring, followed by addition of the amine. No thickening occurs but the cement may appear somewhat lumpy. Lastly, add the water with vigorous mixing, and the cement will immediately thicken to a smooth, easily spread compound. Visc. is 3500 cps (Brookfield, 20 rpm) and Brookfield yield value is 1100.

No. 2

(Solvent-Based)

Hycar® 1432	5.0
Methyl Ethyl Ketone	45.0
Toluene	14.0
Methanol	29.0
Carbopol® 934	2.0
Di-2 (Ethylhexyl) Amine	5.0

No. 3

(Solvent-Based, Nitrile)

Methyl Ethyl Ketone	80.2
Hycar® 1001X225	4.8
Water	11.3
Carbopol® 934	1.9
Di-2 Ethylhexyl Amine	1.9

Procedure:

Dissolve the rubber in the MEK. Add the Carbopol® resin and the amine while agitating. Upon the addition of water, the system will gel to a Brookfield visc. of 3500 cps (20 rpm).

	No. 4	No. 5
	(Vulcanized Natural Rubber)	
Natural Rubber	100.0	100.0
Zinc Oxide	5.0	5.0
Stearic Acid	0.5	0.5
Antioxidant	1.0	1.0
Sulfur	4.0	—
Vulcafor ZIX	—	1.5
Vulcafor DDCN	—	0.5

Vulcanizable Rubber Solution Adhesive

(Solvent-Based)

Natural Rubber (RSS)	10.0
Zinc Oxide	1.0
Antioxidant	0.1
Sulfur	0.1
Solvent	80.0

Procedure:

To 100 parts of the above add 4 parts of a 10% solution of a dithiocarbamate accelerator such as **Butyl 8** and mix thoroughly immediately prior to use.

Hot-Melt Poultry-Curtain Coating

Formula No. 1

(Ethylene Vinyl Acetate/Resin)

Piccopale® 100-SF	10.0–15.0
Elvax 260	10.0–15.0
Paraffin Wax ($\approx$ m.p. 155 F)	70.0–60.0
Microcrystalline Wax ($\approx$ m.p. 180 F)	10.0
Tenox BHT	0.1

No. 2

(Glossy, Ethylene Vinyl Acetate/Resin)

Piccolyte® A-125	27.5
Elvax 250	23.0
Paraffin Wax (≈ m.p. 155 F)	40.0
Micro Wax (≈ m.p. 180 F)	8.0
Advawax 280	1.5
Tenox BHT	0.1

Flame-Retardant Adhesive

(Solvent-Based, Rubber)

Neoprene AC	100.0
Zinc Oxide	5.0
Maglite D	8.0
SP-134	45.0
Antimony Trioxide	50.0
Tricresyl Phosphate	3.0
Ionol	2.0
Methyl Ethyl Ketone	213.0
Hexane	213.0
Toluene	213.0

Cure @ 280 F.

Flameproof Adhesives

Formula No. 1

(PVAc/Resin)

Gelva S-55	85–90
Santicizer 148	15–10

Note:
Where FDA requirements must be met, the use of **Santicizer 141** in place of **Santicizer 148** is recommended.

No. 2

(PVC)

Geon® 577	89
Picco® A-60	91
Good-Rite® K-718	16
Calcium Carbonate, Dispersion A	43
Thermoguard S, Dispersion B (below)	7

Dispersion Component:	A	B
Water	28	28
Nopcosant K	2	2
Calcium Carbonate	70	—
Thermoguard S	—	20

Heat-and-Oil-Resistant Adhesive

(Latex/Resin)

Chemigum N-S	100.0
Hi-Sil 233	10.0
Zinc Oxide	5.0
Sulfur	2.5
2-Mercapto Benzothiazole	1.5
SP-8014	100.0
Methyl Ethyl Ketone	657.0

Nonallergic Bandage Adhesive

(Water-Based)

Zinc Oxide	15
Gelatin	15
Glycerin	25
Water	45

Procedure:

Place the gelatin in water and allow to soften and swell. Add glycerin and heat over a waterbath, stirring gently until solution is effected. Finally incorporate the zinc oxide and stir until cool.

Cigar-Wrapper Repair Adhesive

(Water-Based)

Gelatin	1.0
Hyamine 4000	0.1
Water	99.0

Note:

When a cigar wrapper is broken the above is applied to reattach it and prevent uncurling.

Denture Adhesives

Formula No. 1

(Cellulose Gum)

	Powder	Cream
CMC-7H3SXF		
(or **CMC-7HXF** or **CMC-7H4XF**)	66.0	33.0
Talcum (USP)	33.0	16.0
Perfume and Color	1.0	1.0
Petrolatum	—	50.0

No. 2

(Cellulose Gum)

	Powder	Cream
CMC-4H1F	33.0	17.0
CMC-12M31P	33.0	17.0
Talcum (USP)	33.0	15.0
Perfume and Color	1.0	1.0
Petrolatum	—	50.0

Note:

Plastics such as polyoxyethylene esters, microfine polyethylene, or polypropylene may be added to improve effectiveness of either powder or cream formulations.

Latex Adhesive

(Resin)

Croturez B-115	40.0
Mineral Spirits	13.3
Tall Oil Fatty Acid (high quality)	3.4
Caustic Potash Sol'n.*	5.7
Water	37.5

---

* 6.18 parts of 85% KOH in 93.82 parts of water.

Procedure:

Croturez B-115 is dissolved in the mineral spirits and the fatty acid blended in. The caustic solution is then added with agitation followed by water added at a fairly rapid rate. This emulsion remains stable several days. (Adding water at a very slow rate forms a water-in-oil emulsion.)

Note:

The above emulsion can be blended with a 40% natural rubber latex in different ratios which will produce different adhesive effects. Ratios of 3/1 and 1/1 parts of rubber latex to Croturez emulsion produce films tacky to touch. One part of rubber latex to three parts of Croturez emulsion produce films, when dried, that are nontacky to touch but tacky to each other and suitable for self-sealing films.

Specialty Adhesive

	Formula No. 1	No. 2	No. 3	No. 4
	(Solvent-Based, Resin)			
Ethyl Cellulose N-22	—	60	40	10
Ethyl Cellulose N-100	50	—	—	—
Staybelite® Ester 10	—	—	25	—
Abitol®	—	—	25	—
Lewisol® 28	—	30	—	—
Cumar® P25	—	—	—	45

Uses:

(No. 1): General-purpose; excellent bonding between glass–wood and glass–metal and cardboard–wood.

(No. 2): Adhering ethyl cellulose film or plastic to other surfaces.
(No. 3): Between two filled ethyl cellulose plastic surfaces.
(No. 4): Heat-activated adhesives for labels.

Increased-Rigidity Adhesive

(Polyvinyl Acetate/Latex)

Natural Latex (60%)	167
Zinc Oxide (33% dispersion)	3
Polyvinyl Acetate (50% emulsion)	200
Vulcastab LS (10% sol'n.)	10
Antioxidant (25% sol'n.)	4
Zinc Diethyl Dithiocarbamate (33% dispersion)	3
Sulfur (50% dispersion)	2

Quick-Tack Adhesive

(Polyvinyl Acetate)

Gelva S-55	100
Benzoflex 2-45	12–13
Santicizer 8	12
Water	8–9

Quick-Grab Adhesives

	Formula No. 1	No. 2
	(Latex)	
NR Latex (60%), LA-TZ	167.0	—
HRH Latex* (66%)	—	152.0
Toluene	3.0–5.0	—
Zinc Diethyl Dithiocarbamate (50%)	2.0	2.0

---

* Latex modified with 0.15% hydroxylamine immediately after concentration.

Quick-Break Adhesive

(Latex)

Natural Rubber Latex (60%)	100.00
Coal-Tar Naphtha	10.00
Pale Ester Gum	2.60
Wood Rosin	0.46
Oleic Acid	0.32
Anionic Surfactant	0.50

Nonsag, Elevated-Temperature Cure Encapsulation Compound

(Butyl Rubber)

Enjay Butyl LM 430	100
Mistron Vapor® Talc	70
Whitetex Clay	30
Flexon 845	50
MPC 66 Black	5
Zinc Oxide	5
Maglite K	5
Molecular Sieve 4A	5
Stearic Acid	1
Para-Quinone Dioxime	4
Pb_3O_4	10
Altax	4

Note:

This compound would find application in areas where a nonsag compound is desired for gun or transfer molding application and elevated-temperature cure. Pot life is over 1 h @ 200 F.

Room-Temperature Cure Potting Compounds

		Formula No. 1	No. 2
		Self-leveling	Nonsag
		(Butyl Rubber)	
A	**Enjay Butyl LM 430**	50.00	50.00
	Para-Quinone Dioxime	3.50	3.50
	Toluene	17.00	17.00
B	**Enjay Butyl LM 430**	50.00	50.00
	Mistron Vapor® Talc	—	15.00
	Whitetex Clay	40.00	30.00
	Neodecanoic Acid	0.33	0.33
	Lead Oxide	7.50	7.50
	Toluene	15.00	15.00

Properties: (Compounds mixed by dispersing dry ingredients in polymer/solvent blend on a 3-roll paint mill).

Mix Ratio	1:1.5	1:1.5
Gel Time, h	1.8	1.3
Visc., Brookfield RTV @ 5 rpm, cps		
Part A	129,000	129,000
Part B	190,000	368,000

Note:

These two-part flexible potting compounds can be either poured in place or dispensed by automatic equipment. Through adjustment of curative concentrations of the PbO_2 type, or by the addition of cure activators, a wide variety of pot lives and cure times is possible. It is also possible to adjust the formulations to provide a variation of mix ratios to fit particular application systems.

Two-Part Bonding Adhesives

(Solvent-Based, Butyl Rubber)

A	**Enjay Butyl 218**	100.0
	Staybelite® Ester 10	40.0
	Stearic Acid	3.0
	Zinc Oxide	5.0

Sulfur	1.5
GMF	4.0

Cement—g wt.

Masterbatches	400
Solvent Hexane	1950
Isopropyl Alcohol (92%)	20
Solids, wt. %	17

B **Enjay Butyl 218**	100.0
SRF Black	80.0
Stearic Acid	3.0
Zinc Oxide	5.0
Sulfur	1.5

Cement—g. wt.

Masterbatches	540.0
Solvent Hexane	1950.0
Isopropyl Alcohol (92%)	20.0
Solids, wt. %	21.5

Procedure:

Make each part 30% solids in hexane solvent, adding as part of solvent 5 phr of 92% isopropyl alcohol. Mix 50/50 when ready to use. It has 1–6 h gel time depending on water content of fillers and hexane. The water helps to activate the cure.

Low-Temperature Adhesive

	Formula No. 1	No. 2
	(Latex/Resin)	
Hycar® 4004	100.0	100.0
DOP	10.0	10.0
Piccotex® LC	—	40.0
Ester Gum	90.0	—

Dispersion Adhesives
Formula No. 1
(Water-Based)

Irganox® 1010	50.0
Ammoniated Casein (18% sol'n.)	6.0
Darvan® #7 (25% sol'n. in water)	3.0
Distilled Water	41.0

Procedure:

Dissolve ammoniated casein and **Darvan® #7** in distilled water and add to **Irganox® 1010** with gentle stirring. Subject the resulting mixture to ball milling for about 48 h. Use of porcelain balls is recommended.

Note:

A stable 40–50% w/w **Irganox® 1035** dispersion can be prepared by using the same general procedure as **Irganox® 1076**, but requiring a lesser amount of **Irganox® 1035** (melting point 65–68 C) and a somewhat higher water-phase temperature (70–72 C).

This dispersion, however, is not stable for long periods of time and should be used within 8–24 h max. of preparation. The lower activity (40–50% w/w) of the **Irganox® 1035** dispersion is recommended in order to achieve a flowable or pourable consistency.

No. 2
(Water-Based, Latex)

A Carboxylated SBR Latex	
Carboxylated SBR Latex	
(49% solids; styrene-butadiene ratio 46:54)	100.0
Calcium Carbonate	136.0
Acrysol® GS	1.3
Tetrasodium Pyrophosphate (5% sol'n.)	5.2
Antioxidant*	as shown

---

* Antioxidant

Irganox® 1076	100.00	} Oil Phase—Part A
Oleic Acid (USP)	10.00	
Sodium Hydroxide (98.8% pure)	1.44	} Water Phase—Part B
Distilled Water	50.00	

B Commercially Available SBR
Solvent-Based Pressure-Sensitive Adhesive

C Commercially Available Carboxylated SBR Latex

Procedure:

Melt the **Irganox®** **1076** with 10% of its weight of oleic acid at 60–65 C. The melted mixture of **Irganox®** **1076** and oleic acid (Part A) is added very slowly with vigorous stirring to hot distilled water (60–65 C) containing a stoichiometric amount of sodium hydroxide (Part B). Any high-speed mixer can be employed. Stir continuously for 1–2 min and then quickly cool the mixture using a cold water bath (20–25 C) while continuing stirring for 10–15 min.

The **Irganox®** **1076** dispersion, once properly prepared, is stable against settling or agglomeration for several months provided it is protected from freezing and from exposure to air and light. If it will not be used within 24 h, the dispersion should be stored in amber-colored or nonlight-transmitting containers filled to exclude air. Alternatively, the space above the liquid level maybe filled with nitrogen. Failure to observe these precautions will result in eventual yellowing of the air/dispersion interface. None of the above precautions are required when the dispersion will be used shortly after preparation. Dispersions of lower activity may be prepared by decreasing the proportion of **Irganox®** **1076** in the organic phase.

APPENDIX

pH Values

Acids	pH Value	Bases	pH Value
Hydrochloric Acid	1.0	Sodium Bicarbonate	8.4
Sulfuric Acid	1.2	Borax	9.2
Phosphoric Acid	1.5	Ammonia	11.1
Sulfuric Acid	1.5	Sodium Carbonate	11.6
Acetic Acid	2.9	Trisodium Phosphate	12.0
Alum	3.2	Sodium Metasilicate	12.2
Carbonic Acid	3.8	Lime, Saturated	12.3
Boric Acid	5.2	Sodium Hydroxide	13.0

pH Ranges of Common Indicators

	Useful pH Range
Thymol	1.2 – 2.8
Bromphenol Green	2.8 – 4.6
Methyl Orange	3.1 – 4.4
Bromcresol Green	4.0 – 5.6
Methyl Red	4.4 – 6.0
Propyl Red	4.8 – 6.4
Bromcresol Purple	5.2 – 6.8
Brom Thymol Blue	6.0 – 7.6
Phenol Red	6.8 – 8.4
Litmus	7.2 – 8.8
Cresol Red	7.2 – 8.8
Cresolphthalein	8.2 – 9.8
Phenolphthalein	8.6 – 10.2
Nitro Yellow	10.0 – 11.6
Alizarin Yellow R	10.1 – 12.1
Sulfo Orange	11.2 – 12.6

International Atomic Weights

Element	Symbol	Atomic Number	Atomic Weight	Element	Symbol	Atomic Number	Atomic Weight
Actinium	Ac	89	(227)	Chlorine	Cl	17	35.453
Aluminum	Al	13	26.9815	Chromium	Cr	24	51.996
Americium	Am	95	(243)	Cobalt	Co	27	58.9332
Antimony	Sb	51	121.75	Copper	Cu	29	63.54
Argon	Ar	18	39.948	Curium	Cm	96	(247)
Arsenic	As	33	74.9216	Dysprosium	Dy	66	162.50
Astatine	At	85	(210)	Einsteinium	Es	99	(254)
Barium	Ba	56	137.34	Element 102		102	(254)
Berkelium	Bk	97	(247)	Erbium	Er	68	167.26
Beryllium	Be	4	9.0122	Europium	Eu	63	151.96
Bismuth	Bi	83	208.980	Fermium	Fm	100	(253)
Boron	B	5	10.811	Fluorine	F	9	18.9984
Bromine	Br	35	79.909	Francium	Fr	87	(223)
Cadmium	Cd	48	112.40	Gadolinium	Gd	64	157.25
Calcium	Ca	20	40.08	Gallium	Ga	31	69.72
Californium	Cf	98	(251)	Germanium	Ge	32	72.59
Carbon	C	6	12.01115	Gold	Au	79	196.967
Cerium	Ce	58	140.12	Hafnium	Hf	72	178.49
Cesium	Cs	55	132.905	Helium	He	2	4.0026

Element	Symbol	Atomic Number	Atomic Weight
Holmium	Ho	67	164.930
Hydrogen	H	1	1.00797
Indium	In	49	114.82
Iodine	I	53	126.9044
Iridium	Ir	77	192.2
Iron	Fe	26	55.847
Krypton	Kr	36	83.80
Lanthanum	La	57	138.91
Lawrencium	Lw	103	(256)
Lead	Pb	82	207.19
Lithium	Li	3	6.939
Lutetium	Lu	71	174.97
Magnesium	Mg	12	24.312
Manganese	Mn	25	54.9380
Mendelevium	Md	101	(256)
Mercury	Hg	80	200.59
Molybdenum	Mo	42	95.94
Neodymium	Nd	60	144.24
Neon	Ne	10	20.183

Element	Symbol	Atomic Number	Atomic Weight
Neptunium	Np	93	(237)
Nickel	Ni	28	58.71
Niobium	Nb	41	92.906
Nitrogen	N	7	14.0067
Osmium	Os	76	190.2
Oxygen	O	8	15.9994
Palladium	Pd	46	106.4
Phosphorus	P	15	30.9738
Platinum	Pt	78	195.09
Plutonium	Pu	94	(244)
Polonium	Po	84	(209)
Potassium	K	19	39.102
Praseodymium	Pr	59	140.907
Promethium	Pm	61	(145)
Protactinium	Pa	91	(231)
Radium	Ra	88	(226)
Radon	Rn	86	(222)
Rhenium	Re	75	186.2
Rhodium	Rh	45	102.905

International Atomic Weights (cont'd.)

Element	Symbol	Atomic Number	Atomic Weight	Element	Symbol	Atomic Number	Atomic Weight
Rubidium	Rb	37	85.47	Terbium	Tb	65	158.924
Ruthenium	Ru	44	101.07	Thallium	Tl	81	204.37
Samarium	Sm	62	150.35	Thorium	Th	90	232.038
Scandium	Sc	21	44.956	Thulium	Tm	69	168.934
Selenium	Se	34	78.96	Tin	Sn	50	118.69
Silicon	Si	14	28.086	Titanium	Ti	22	47.90
Silver	Ag	47	107.870	Tungsten	W	74	183.85
Sodium	Na	11	22.9898	Uranium	U	92	238.08
Strontium	Sr	38	87.62	Vanadium	V	23	50.942
Sulfur	S	16	32.064	Xenon	Xe	54	131.30
Tantalum	Ta	73	180.948	Ytterbium	Yb	70	173.04
Technetium	Tc	43	(97)	Yttrium	Y	39	88.905
Tellurium	Te	52	127.60	Zinc	Zn	30	65.37
				Zirconium	Zr	40	91.22

Temperature Conversion Tables

F	C	F	C	F	C	F	C	F	C
−40	−40.0	−2	−18.9	35	1.7	72	22.2	109	42.8
−39	−39.4	−1	−18.3	36	2.2	73	22.8	110	43.3
−38	−38.9	0	−17.8	37	2.8	74	23.3	111	43.9
−37	−38.3	1	−17.2	38	3.3	75	23.9	112	44.4
−36	−37.8	2	−16.7	39	3.9	76	24.4	113	45.0
−35	−37.2	3	−16.1	40	4.4	77	25.0	114	45.6
−33	−36.1	4	−15.6	41	5.0	78	25.6	115	46.1
−32	−35.6	5	−15.0	42	5.6	79	26.1	116	46.7
−31	−35.0	6	−14.4	43	6.1	80	26.7	117	47.2
−30	−34.4	7	−13.9	44	6.7	81	27.2	118	47.8
−29	−33.3	8	−13.3	45	7.2	82	27.8	119	48.3
−28	−33.3	9	−12.8	46	7.8	83	28.3	120	48.9
−27	−32.8	10	−12.2	47	8.3	84	28.9	121	49.4
−26	−32.2	11	−11.7	48	8.9	85	29.4	122	50.0
−25	−31.7	12	−11.1	49	9.4	86	30.0	123	50.6
−24	−31.1	13	−10.6	50	10.0	87	30.6	124	51.1
−23	−30.6	14	−10.0	51	10.6	88	31.1	125	51.7
−22	−30.0	15	−9.4	52	11.1	89	31.7	126	52.2
−21	−29.4	16	−8.9	53	11.7	90	32.2	127	52.8
−20	−28.9	17	−8.3	54	12.2	91	32.8	128	53.3
−19	−28.3	18	−7.8	55	12.8	92	33.3	129	53.9
−18	−27.8	19	−7.2	56	13.3	93	33.9	130	54.4
−17	−27.2	20	−6.7	57	13.9	94	34.4	131	55.0
−16	−26.7	21	−6.1	58	14.4	95	35.0	132	55.6
−15	−26.1	22	−5.6	59	15.0	96	35.6	133	56.1
−14	−25.6	23	−5.0	60	15.6	97	36.1	134	56.7
−13	−25.0	24	−4.4	61	16.1	98	36.7	135	57.2
−12	−24.4	25	−3.9	62	16.7	99	37.2	136	57.8
−11	−23.9	26	−3.3	63	17.2	100	37.8	137	58.3
−10	−23.3	27	−2.8	64	17.8	101	38.3	138	58.9
−9	−22.8	28	−2.2	65	18.3	102	38.9	139	59.4
−8	−22.2	29	1.7	66	18.9	103	39.4	140	60.0
−7	−21.7	30	−1.1	67	19.4	104	40.0	141	60.6
−6	−21.1	31	−0.6	68	20.0	105	40.6	142	61.1
−5	−20.6	32	0.0	69	20.6	106	41.1	143	61.7
−4	−20.0	33	0.6	70	21.1	107	41.7	144	62.2
−3	−19.4	34	1.1	71	21.7	108	42.2	145	62.8

Temperature Conversion Tables *(cont'd.)*

F	C	F	C	F	C	F	C	F	C
146	63.3	183	83.9	220	104.4	257	125.0	294	145.6
147	63.9	184	84.4	221	105.0	258	125.6	295	146.1
148	64.4	185	85.0	222	105.6	259	126.1	296	146.7
149	65.0	186	85.6	223	106.1	260	126.7	297	147.2
150	65.6	187	86.1	224	106.7	261	127.2	298	147.8
151	66.1	188	86.7	225	107.2	262	127.8	299	148.3
152	66.7	189	87.2	226	107.8	263	128.3	300	148.9
153	67.2	190	87.8	227	108.3	264	128.9	301	149.4
154	67.8	191	88.3	228	108.9	265	129.4	302	150.0
155	68.3	192	88.9	229	109.4	266	130.0	303	150.6
156	68.9	193	89.4	230	110.0	267	130.6	304	151.1
157	69.4	194	90.0	231	110.6	268	131.1	305	151.7
158	70.0	195	90.6	232	111.1	269	131.7	306	152.2
159	70.6	196	91.1	233	111.7	270	132.2	307	152.8
160	71.1	197	91.7	234	112.2	271	132.8	308	153.3
161	71.7	198	92.2	235	112.8	272	133.3	309	153.9
162	72.2	199	92.8	236	113.3	273	133.9	310	154.4
173	72.8	200	93.3	237	113.9	274	134.4	311	155.0
164	73.3	201	93.9	238	114.4	275	135.0	312	155.6
165	73.9	202	94.4	239	115.0	276	135.6	313	156.1
166	74.4	203	95.0	240	115.6	277	136.1	314	156.7
167	75.0	204	95.6	241	116.1	278	136.7	315	157.2
168	75.6	205	96.1	242	116.7	279	137.2	316	157.8
169	76.1	206	96.7	243	117.2	280	137.8	317	158.3
170	76.7	207	97.2	244	117.8	281	138.3	318	158.9
171	77.2	208	97.8	245	118.3	282	138.9	319	159.4
172	77.8	209	98.3	246	118.9	283	139.4	320	160.0
173	78.3	210	98.9	247	119.4	284	140.0	321	160.6
174	78.9	211	99.4	248	120.0	285	140.6	322	161.1
175	79.4	212	100.0	249	120.6	286	141.1	323	161.7
176	80.0	213	100.6	250	121.1	287	141.7	324	162.2
177	80.6	214	101.1	251	121.7	288	142.2	325	162.8
178	81.1	215	101.7	252	122.2	289	142.8	326	163.3
179	81.7	216	102.2	253	122.8	290	143.3	327	163.9
180	82.2	217	102.8	254	123.3	291	143.9	328	164.4
181	82.8	218	103.3	255	123.9	292	144.4	329	165.0
182	83.3	219	103.9	256	124.4	293	145.0	330	165.6

Temperature Conversion Tables *(cont'd.)*

F	C	F	C	F	C	F	C	F	C
331	166.1	355	179.4	380	193.3	404	206.7	428	220.0
332	166.7	356	180.0	381	193.9	405	207.2	429	220.6
333	167.2	357	180.6	382	194.4	406	207.8	430	221.1
334	167.8	359	181.7	383	195.0	407	208.3	431	221.7
335	168.3	360	182.2	384	195.6	408	208.9	432	222.2
336	168.9	361	182.8	385	196.1	409	209.4	433	222.8
337	169.4	362	183.3	386	196.7	410	210.0	434	223.3
338	170.0	363	183.9	387	197.2	411	210.6	435	223.9
339	170.6	364	184.4	388	197.8	512	211.1	436	224.4
340	171.1	365	185.0	389	198.3	413	211.7	437	225.0
341	171.7	366	185.6	390	198.9	414	212.2	438	225.6
342	172.2	367	186.1	391	199.4	415	212.8	439	226.1
343	172.8	368	1186.7	392	200.0	416	213.3	440	226.7
344	173.3	369	187.2	393	200.6	417	213.9	441	227.2
345	173.9	370	187.8	394	201.1	418	214.4	442	227.8
346	174.4	371	188.3	395	201.7	419	215.0	443	228.3
347	175.0	372	188.9	396	202.2	420	215.6	444	228.9
348	175.0	373	189.4	397	202.8	421	216.1	445	229.4
349	176.1	374	190.0	398	203.3	422	216.7	446	230.0
350	176.7	375	190.6	399	203.9	423	217.2	447	230.6
351	177.2	376	191.1	400	204.4	424	217.8	448	231.1
352	177.8	377	191.7	401	205.0	425	218.3	449	231.7
353	178.3	378	192.2	402	205.6	426	218.9		
354	178.9	379	192.8	403	206.1	427	219.4		

Some Incompatible Chemicals

The substances in the lefthand column must be stored and handled so that they cannot come into contact with the substances in the righthand column.

Acetic acid

Chromic acid, nitric acid, hydroxyl-containing compounds, ethylene glycol, perchloric acid, peroxides, and permanganates.

Acetone	Concentrate nitric and sulfuric acid mixtures.
Acetylene	Chlorine, bromine, copper, silver, fluorine, and mercury.
Alkaline and alkaline-earth metals, such as sodium, potassium, cesium, lithium, magnesium, calcium, aluminum	Carbon dioxide, carbon tetrachloride, and other chlorinated hydrocarbons. (Also prohibit water, foam, and dry chemical on fires involving these metals.)
Ammonia (anhydr.)	Mercury, chlorine, calcium hypochlorite, iodine, bromine, and hydrogen fluoride.
Ammonium nitrate	Acids, metal powders, flammable liquids, chlorates, nitrites, sulfur, finely divided organics or combustibles.
Aniline	Nitric acid, hydrogen peroxide.
Bromine	Ammonia, acetylene, butadiene, butane and other petroleum gases, sodium carbide, turpentine, benzene, and finely divided metals.
Calcium carbide	Water (See also acetylene).
Calcium oxide	Water.
Carbon, activated	Calcium hypochlorite.
Chlorates	Ammonium salts, acids, metal powders, sulfur, finely divided organics or combustibles.
Chlorine	Ammonia, acetylene, butadiene, butane and other petroleum gases, hydrogen, sodium carbide, turpentine, benzene, and finely divided metals.
Chlorine dioxide	Ammonia, methane, phosphine, and hydrogen sulfide.
Chromic acid	Acetic acid, naphthalene, camphor, glycerol, turpentine, alcohol, and other flammable liquids.
Copper	Acetylene, hydrogen peroxide.
Fluorine	Isolate from everything.

Hydrocarbons (benzene, butane, propane, gasoline, turpentine, etc.)	Fluorine, chlorine, bromine, chromic acid, sodium peroxide.
Hydrocyanic acid	Nitric acid, alkalis.
Hydrofluoric acid, anhydrous (hydrogen fluoride)	Aqueous or anhydrous ammonia.
Hydrogen peroxide	Copper, chromium, iron, most metals or their salts, any flammable liquid, combustible aniline, nitro-methane.
Hydrogen sulfide	Fuming nitric acid, oxidizing gases.
Iodine	Acetylene, anhydrous or aqueous ammonia.
Mercury	Acetylene, fulminic acid, ammonia.
Nitric acid (conc.)	Acetic acid, aniline, chromic acid, hydrocyanic acid, hydrogen sulfide, flammable liquids, flammable gases, and nitritable substances.
Nitroparaffins	Inorganic bases.
Oxalic acid	Silver, mercury.
Oxygen	Oils, grease, hydrogen, flammable liquids, solids or gases.
Perchloric acid	Acetic anhydride, bismuth and its alloys, alcohol, paper, wood, grease, oils.
Peroxides, organic	Organic or mineral acids; avoid friction.
Phosphorus (white)	Air, oxygen.
Potassium chlorate	Acids (See also chlorate.)
Potassium perchlorates	Acids (See also perchloric acid.)
Potassium permanganate	Glycerol, ethylene glycol, benzaldehyde, sulfuric acid.
Silver	Acetylene, oxalic acid, tartaric acid, fulminic acid, ammonium compounds.
Sodium	See alkaline metals.
Sodium nitrate	Ammonium nitrate and other ammonium salts.

Sodium oxide	Water.
Sodium peroxide	Any oxidizable substances, such as ethanol, methanol, glacial acetic acid, acetic anhydride, benzaldehyde, carbon disulfide, glycerol, ethylene glycol, ethyl acetate, methyl acetate, and furfural.
Sulfuric acid	Chlorates, perchlorates, permanganates.
Zirconium	Prohibit water, carbon tetrachloride, foam, and dry chemical on zirconium fires.

Safety in the Laboratory or Home Workshop

It is necessary to learn:
Use of laboratory fume hoods
Handling flammable solvents
Mixing acids
Glass blowing

Common electrical hazards
First aid (for four situations only)
Stoppage of breathing
Profuse bleeding
Chemical burns (water only)
Fire in clothing
Use of portable fire extinguishers
Compressed or flammable gases
Handling and storing dangerous chemicals, including alkali metals.

An outstanding deficiency pertaining to laboratory safety seems to be a lack of awareness of hazards among nontechnical personnel. It is conceivable that increased emphasis on "briefing" custodial workers about the dangers of the laboratories in which they work, and periodic review of these conditions could substantially reduce the hazard of ignorance.

A more universal use of safety glasses, reaction shields, and other personal protective devices seems to be needed. From the responses received, an increased program of education on the hazards of common laboratory procedures and the use of personal protective equipment to lessen these hazards would be helpful.

Chemical Hazards

All laboratories, whether they be biological, chemical, or radiological, utilize hazardous chemicals. The hazard may result from utilizing the "raw" product or from products of a chemical reaction between two or more substances or breakdown products developed through heating or aging. Laboratory personnel should have an acquaintance at least with the modes of entry, the physiological responses, both acute and chronic, and methods of roughly assessing the hazards potential of chemicals they are using.

Electrial Hazards and Management

The problem of handling electricity is probably one of the most ignored facets of safety, yet each year many needless deaths and injuries are caused through carelessness in handling even low voltages. It is also of importance to recognize that electrical equipment can act as an ignition source to activate a fire or explosion. Static electricity should be considered in this category.

Pressure Hazards

Pressure equipment, either high or low (vacuum), is a part of most laboratories. High-pressure apparatus such as gas cylinders, if improperly handled, can be very dangerous. This is especially true of oxygen. Precautions are necessary in handling, transporting, and in storing. Vacuum equipment, through implosions, can be every bit as dangerous as high-pressure explosions.

Cryogenic Hazards

Cryogenics or the use of low-temperature refrigerants requires a knowledge of the behavior of these materials under laboratory conditions. It is impossible to understand the design of a piece of cryogenic equipment or cryogenic experiment without an appreciation for the principles of insulation or the significance of extremely low temperatures. Misuse can result in severe injury.

Flammable Chemicals-Hazard

Fires and explosions account for the most dangerous and the most expensive types of laboratory accidents. A knowledge of the flammable properties of chemicals along with an understanding of potential sources of ignition is extremely vital. Storage and handling of these materials also requires special attention.

General Safety Considerations

A number of accidents and injuries in laboratories could very well result from improper lifting, falls, and lacerations from improper handling of glassware. Preventive measures in these areas are worthy of mention.

Ventilation

The principal method of hazards control in laboratory involves the effective use of ventilation, both general and exhaust. An example of exhaust ventilation is the fume hood which if improperly designed or used fails to give the desired protection. Observations indicate that the function of this equipment is not entirely understood and a number of misuses have been witnessed.

Laboratory Sanitation

Poor laboratory sanitation practices may be the cause of contaminating potable water supplies through temporary cross connections. At times, poor housekeeping practices may be dangerous because of blocking passages or by providing tripping hazards. Many chemicals are kept well beyond their usefulness, causing containers to deteriorate and leak, or chemicals to become unstable. Disposal of flammable and toxic chemicals also presents a problem.

Protective Equipment

All laboratories require that protective equipment of one type or another be immediately available. These devices may include eye wash, emergency shower, safety glasses, eye shields, protective clothing, and respiratory protection. Knowledge of the proper usage and limitations of such equipment is extremely important. At times, injury or death may result from improper selection and application of protective equipment.

Reports and Records

Reports and records are necessary adjuncts to any safety program and should be complete, accurate, and disseminated to the appropriate administrators. Accident reports are of little value unless periodically examined and tabulated in order to obtain a picture of local and overall problems.

Emergencies

The initial procedures one follows in an emergency often determine the ultimate outcome of the accident, both to the individuals and to the installation. The rudiments of first aid, fire fighting, and reporting are vital. Personnel have to be continually instructed on procedures for medical and file emergencies and how and where to make these initial contacts. Such procedures are critical, especially when working alone.

Contact Lenses

It is important to wear eye protection in the chemical laboratory and especially when wearing contact lenses.

Danger in Handling Acid

The heat evolution of the solution could cause sufficient thermal shock to a glass container to permit it to crack when lifted free of a counter or set on a cold counter top; or a shock in setting it down could contribute to the bottom separating when the glass container is lifted.

Written procedures for handling of acids should always be followed. Personal protective equipment consisting of face protection, rubber apron, and gloves are a necessity for this operation.

You Must Have Fire Extinguishers

If a fire breaks out in your office or apartment, get out fast. Many people are killed because they don't realize how fast a small fire can spread.

If you are caught in smoke take short breaths, breathe through your nose, and crawl to escape. The air is better near the floor.

Head for stairs—not elevator. A bad fire can cut off the power to elevators. Close all doors and windows behind you.

If you are trapped in a smoke-filled room, stay near the floor, where the air is better. If possible, sit by a window where you can call for help.

Feel every door with your hand. If it's hot don't open. If it's cool, make this test: open slowly and stay behind the door. If you feel heat or pressure coming through the open door, slam it shut.

If you can't get out, stay behind a closed door. Any door serves as a shield. Pick a room with a window. Open the window at the top and bottom. Heat and smoke will go out the top. You can breathe out the bottom.

DON'T fight a fire yourself.

DON'T jump. Many people have jumped and died without realizing rescue was just a few minutes away.

If there is a panic for the main exit, get away from the mob. Try to find another way out. Once you are safely out, *DON'T* go back in. Call the Fire Department immediately. Use alarm box or telephone.

If you find smoke in an open stairway or open hall, use another preplanned way out.

REMEMBER: Get out fast. Don't underestimate how fast a small fire can spread. Use stairs, not the elevator. Close all doors behind you. Don't panic. Once you are safely out, call the Fire Department. Use alarm box. Don't go back in.

Trademark Chemical Manufacturers

The following is a list of Trademark Chemical Manufacturers. The names of these manufacturers are preceded by a number. In the list of Trademark Chemicals that follows this one, the manufacturers are referred to by a number alone.

1	ABM Chem.	Leeds, UK
2	Aceto Chem. Co.	Flushing, NY
3	Air Prod. & Chem.	Allentown, PA
4	Akzo Chemie.	Burt, NY
5	Alcolac Inc.	Baltimore, MD
6	Allied Chem. Corp.	Morristown, NJ
7	Aluminum Co. of Amer.	Pittsburgh, PA
8	American Colloid Co.	Skokie, IL
9	American Cyanamid.	Wayne, NJ
10	American Hoechst Corp.	Somerville, NJ
11	Ameripol Inc.	Cleveland, OH
12	Amoco Chem. Corp.	Chicago, IL
13	Arakawa.	Japan
14	Archer-Daniels-Midland Co.	Minneapolis, MN
15	Argus Chem. Corp.	Brooklyn, NY
16	Arkansas Co.	Newark, NJ
17	Arizona Chem. Co.	New York, NY
18	Armak Ind. Chem.	Asheville, NC
19	Associated Lead.	London, UK
20	Atlantic Powdered Metals.	New York, NY
21	Atlantic Richfield Co.	Philadelphia, PA
22	Bareco Div., Petrolite.	Tulsa, OK
23	BASF AG.	Ludwigshafen, Germany
24	BASF Corp.	Paramus, NJ
25	BASF-Wyandotte.	Wyandotte, MI
26	Bayer AG.	Leverkusen, W. Germany
27	Borden, Inc.	Columbus, OH
28	Borg-Warner Chem., Inc.	Parkersburg, WV
29	British Celanese Ltd.	Coventry, UK
30	Brown Co., R.J.	St. Louis, MO

31	Burgess Pigment Co.	Macon, GA
32	Cabot Corp.	Tuscalo, IL
33	Calgon Corp.	Pittsburgh, PA
34	Carlisle Chem. Works	New Brunswick, NJ
35	Carson Chem.	Long Beach, CA
36	Chem Mark Inc.	Middlesex, NJ
37	Ciba-Geigy Corp.	Greensboro, NC
38	Cities Service Co.	Atlanta, GA
39	Colloids Inc.	Newark, NJ
40	Columbian Div., Cities Service	Atlanta, GA
41	Crosby Chem. Inc.	DeRidder, LA
42	Cyprus Ind. Min. Co.	Los Angeles, CA
43	Degussa, Inc.	Teterboro, NJ
44	Dexter Chem. Corp.	Bronx, NY
45	Diamond Alkali Co.	Cleveland, OH
46	Diamond Shamrock Corp.	Cleveland, OH
47	Douglas Labs.	Chicago, IL
48	Dow Chem. Co.	Midland, MI
49	Dow Corning Corp.	Midland, MI
50	E.I. DuPont	Wilmington, DE
51	Eastman Chem. Prod. Inc.	Kingsport, TN
52	Emery Ind. Inc.	Cincinnati, OH
53	Engelhard Min. & Chem. Corp.	Edison, NJ
54	Enjay Chem. Co.(div. Humble)	New York, NY
55	Ethyl Corp.	Baton Rouge, LA
56	Exxon Corp.	Houston, TX
57	FMC Corp.	Philadelphia, PA
58	Freeport Kaolin.	Gordon, GA
59	GAF Corp.	New York, NY
60	General Mills Chem. Inc.	Minneapolis, MN
61	Genstar Stone Prod.	Hunt Valley, MD
62	Georgia Kaolin Co.	Elizabeth, NJ
63	Georgia Marble.	Atlanta, GA
64	Georgia-Pacific.	Portland, OR
65	GE, Polymers Prod. Dept.	Pittsfield, MA
66	Germains Inc.	Los Angeles, CA
67	Glidden Durkee Div. SCM Corp.	Jacksonville, FL
68	Globe Albany Corp.	Albany, NY
69	B.F. Goodrich Chem. Co.	Cleveland, OH
70	Goodyear Tire & Rubber Co.	Akron, OH

71	W.R. Grace & Co...Lexington, MA
72	Great Lakes Chem. Corp..................................W. Lafayette, IN
73	Gulf Oil Chem. Co...Houston, TX
74	The C.P. Hall Co...Chicago, IL
75	Hardman Inc...Belleville, NJ
76	Harshaw Chem. Co..Cleveland, OH
77	Harwick...Akron, OH
78	Hercules Inc..Wilmington, DE
79	Hooker Chem. & Plastics Corp.......................Niagara Falls, NY
80	J.M. Huber Corp...Macon, GA
81	Hydrolabs Inc...Paterson, NJ
82	ICI Americas Inc..Wilmington, DE
83	Industrial Minerals & Chem............................Terre Haute, IN
84	Inmont Corp..Clifton, NJ
85	International Talc Co...New York, NY
86	International Wax Refining Co.......................Valley Stream, NY
87	Johns-Manville Co..Denver, CO
88	Kelco/Div. Merck & Co.....................................San Diego, CA
89	Koppers Co...Pittsburgh, PA
90	KSH Chem. Corp...Kooga/dzaan, Holland
91	Lubrizol Corp...Cleveland, OH
92	Lucidol Corp./Pennwalt Div.................................Buffalo, NY
93	3M Co..St. Paul, MN
94	Merck & Co., Inc..Rahway, NJ
95	Mobay Chem. Corp..Pittsburgh, PA
96	Mobil Chem. Co...Richmond, VA
97	Monsanto Co..St. Louis, MO
98	Moore & Munger..New York, NY
99	M&T Chemicals Inc...Rahway, NJ
100	Nalco Chem. Co..Chicago, IL
101	National Lead Co...New York, NY
102	Neville Chem. Co..Pittsburgh, PA
103	Newport Div..Pensacola, FL
104	NJ Zinc Co..Palmerton, PA
105	NL Industries...Hightstown, NJ
106	Norwegian Talc Inc...England
107	Omya Inc..Proctor, VT
108	Para-Chem Inc...Philadelphia, PA
109	Penick & Ford Inc...Cedar Rapids, IA
110	Pennwalt Corp..Philadelphia, PA

111	Petrolite Corp.	New York, NY
112	Pfizer Inc.	Greensboro, NC
113	Phillips Petroleum Co.	Bartlesville, OK
114	Polymer Applications, Inc.	Tonawanda, NY
115	Polymer Systems Corp.	Pittsburgh, PA
116	Polysar Ltd.	Sarnia, Canada
117	Polyvinyl Chem. Ind.	Wilmington, MA
118	PPG Industries Inc.	Pittsburgh, PA
119	PQ Corp.	Valley Forge, PA
120	Quaker Oats Co.	Chicago, IL
121	Ralston Purina Co.	St. Louis, MO
122	Reichhold Chem.	White Plains, NY
123	Rewo Chem. Co.	Bohemia, NY
124	Rohm & Haas	Philadelphia, PA
125	H.M. Royal, Inc.	Trenton, NJ
126	Sartomer Co.	West Chester, PA
127	Schenectady Chem. Inc.	Schenectady, NY
128	Scott Bader & Co. Ltd.	Northamptonshire, UK
129	Shell Chem. Co.	Houston, TX
130	Sierra Talc & Chem. Corp.	S. Pasadena, CA
131	Skelly Oil Inc.	Tulsa, OK
132	Soltex Polymer Corp.	Deer Park, TX
133	Sonneborn Sons Co.	New York, NY
134	A.E. Staley Mfg. Co.	Decatur, IL
135	Standard Oil	Cleveland, OH
136	Sun Chem. Corp.	Chester, SC
137	Sun Petroleum	Philadelphia, PA
138	SWS Silicones	Adrian, MI
139	Sylacauga Calcium Prod. Co.	Sylacauga, AL
140	Tamms Ind.	Chicago, IL
141	Tenneco Chem.	Piscataway, NJ
142	Texaco Chem. Co.	Austin, TX
143	Thiokol Corp.	Trenton, NJ
144	Thompson & Weinman	Cartersville, GA
145	Troy Chem. Co.	Newark, NJ
146	Ultra Adhesives, Inc.	Paterson, NJ
147	Union Carbide Corp.	New York, NY
148	Union Oil Co.	Los Angeles, CA
149	Uniroyal Chem.	Naugatuck, CT
150	United Clay Mines Corp.	Trenton, NJ

151 U.S. Industrial Chem. Co...................................New York, NY
152 Vanderbilt, R.T. Co., Inc.......................................Norwalk, CT
153 Velsicol Chem. Corp...Chicago, IL
154 Verona Dyestuffs Div...Union, NJ
155 Vulnax International Ltd...................................Manchester, UK
156 Wacker Chemie GmbH...........................Munchen, W. Germany
157 Western Petro Chem Inc...............................Princeton Jct., NJ
158 Wilmington Chem. Corp................................Wilmington, DE
159 Witco Chem..New York, NY

Trademark Chemicals

INDEX